About Island Press

Since 1984, the nonprofit organization Island Press has been stimulating, shaping, and communicating ideas that are essential for solving environmental problems worldwide. With more than 1,000 titles in print and some 30 new releases each year, we are the nation's leading publisher on environmental issues. We identify innovative thinkers and emerging trends in the environmental field. We work with world-renowned experts and authors to develop cross-disciplinary solutions to environmental challenges.

Island Press designs and executes educational campaigns, in conjunction with our authors, to communicate their critical messages in print, in person, and online using the latest technologies, innovative programs, and the media. Our goal is to reach targeted audiences—scientists, policy makers, environmental advocates, urban planners, the media, and concerned citizens—with information that can be used to create the framework for long-term ecological health and human well-being.

Island Press gratefully acknowledges major support from The Bobolink Foundation, Caldera Foundation, The Curtis and Edith Munson Foundation, The Forrest C. and Frances H. Lattner Foundation, The JPB Foundation, The Kresge Foundation, The Summit Charitable Foundation, Inc., and many other generous organizations and individuals.

The opinions expressed in this book are those of the author(s) and do not necessarily reflect the views of our supporters.

True Roots

WHAT QUITTING HAIR DYE TAUGHT ME ABOUT HEALTH AND BEAUTY

Ronnie Citron-Fink

ISLANDPRESS

Washington | Covelo | London

Library of Congress Control Number: 2018967127

All Island Press books are printed on
environmentally responsible materials.

Manufactured in the United States of America
10 9 8 7 6 5 4 3 2 1

Keywords: Aveda, Black Women for Wellness, blending, Brazilian
Blowout, carcinogen, Clairol, coal-tar dyes, colorist, Cosmetic
Ingredient Review, cosmetics, endocrine disruptor, Environmental
Defense Fund, environmental health, Environmental Working Group
(EWG) Skin Deep, Food and Drug Administration (FDA), gray
hair, hair color, hairdresser, Hairprint, Henna, Just For Men, L'Oréal,
Moms Clean Air Force, para-phenylenediamine (PPD), permanent
hair dye, Personal Care Products Council, Pia Grønning, purple
shampoo, salon, semipermanent hair dye, silicon, silver hair, skunk
line, Women's Voices for the Earth (WVE), Yazemeenah Rossi

—for Ted, Lainey, and Ben

Contents

Foreword

Dominique Browning

When I was quite young, my stylish mother taught me to describe her hair color as "salt-and-pepper." I found this an appealing idea, and somehow it made me think everyone's hair color bore some relationship to food: hair the color of strawberries, cinnamon, oats, or—well, if I had ever eaten such a thing, I would have seized on the idea of hair the color of jet-black squid ink. As this is a book about hair and what it is costing us, in health and in dollars, to maintain its splendiferous colors, it is only natural that mothers should march right in at the beginning. It is a universal truth that mothers have an outsize influence in the way we assess and appreciate (or denigrate) ourselves.

My mother's hair began to turn gray when she was in high school, she told me, so I only ever knew her as sporting that dramatic color. The "salt" was bright silver, and the "pepper" was ink black. I recall people stopping her to ask what brand of dye she was using. To which, with a haughty sniff, she would reply, "Nothing." Not that my mother was a naturalist. I don't think I ever saw her without bright red lipstick. Throughout our lives together, my own hair was among her many vexations with my sense of style. By the time I was a teenager, it was so long and unruly that decades later, my own son, watching home movies of me and my sisters, exclaimed in wondrous horror that I looked like a cavewoman.

All of which is to say: hair carries a lot of messages, sends a lot of signals. We see hair before we look into anyone's eyes. When I first met Ronnie Citron-Fink, ten years ago, her hair was the jet black of squid ink, long, shiny, lustrous, and thick. She kept it dark with dye, through painstaking hours sitting in a hairdresser's chair.

Ronnie had just joined me to help launch Moms Clean Air Force. We have become an organization of over a million moms (and probably lots of dads, and aunts, and grandparents too) uniting to demand clean air and climate safety for our families and for the sake of our children's health.

One of the campaigns we focused on in our first few years was the reformation of an outdated law governing chemical safety in this country. Actually, "outdated" doesn't begin to describe the shortcomings of the way the United States reviews chemicals before they get into the stuff we eat, drink, breathe, and slather over our bodies, day after day after day. That's the rub. We have systems to test small doses of single chemicals; we don't have systems to test what happens over time, when a possible carcinogen is applied daily, monthly, yearly. We don't have systems to test what happens when individual chemicals interact with one another as they are absorbed into our bloodstreams.

We are the guinea pigs for the chemical industry.

We think that because we are buying something at a store we trust, it must be safe. That if it doesn't immediately make us sick, or give us a rash, it must be okay. That our government agencies are keeping watch over us—in the case of cosmetics and hair dyes, that would be the Food and Drug Administration (FDA). We are mistaken, sadly. The beauty industry is gigantic: $70 billion per year. The agency that regulates what you put on your hair to color it, and the cosmetics you put on your face to color it: tiny. And dangerously understaffed, with twenty-seven people and an $8 million budget. Nowadays, as Ronnie makes clear, the FDA, more than ever before, is mostly captive to the powerful chemical industry. Day after day, year after year, women, men, children, our babies, even, are exposed to carcinogens, endocrine disrupters, allergens. It is almost impossible to believe, but it is true.

And our hair says it all.

Ronnie and I sat together one afternoon in a meeting about toxic chemicals and the limitations of regulatory agencies. She was, as always, running her fingers through her hair. And then I noticed, suddenly, she stopped. And looked over at me, her still jet black eyebrows high, her gaze critical—and horrified. I could tell what she was thinking: what am I doing to my health by dyeing my hair? Thus began her journey—I would call it an adventure—through what turns out to be the head-spinning maze of loopholes, dark corners, and hidden traps of a multibillion-dollar-per-year hair care industry.

There is a reason, she reports, that hairdressers are three times more likely to get breast cancer, five times more likely to get bladder cancer; why they are much more likely than the average population to have lung and larynx cancers as well as multiple myeloma. That reason lies in the tongue-twisting, if not unpronounceable, strings of chemicals in the dyes we use to maintain hair color. And that's just the chemicals that are listed on the tubes and bottles cluttering our bathroom shelves, the bottles and tubes we reach for, day after day, month after month, year after year.

Another point Ronnie makes is that cosmetics manufacturers don't even have to tell us what's in the stuff they sell us—as one employee of a law firm that, by a weird coincidence, specializes in asbestos litigation, learned when she had her child's makeup tested after her young daughter broke out in a rash while playing with it. She found that the makeup contained asbestos.

Never mind "natural." For the most part, even so-called natural dyes—the rinses and tints and gentle formulas—contain what one activist calls a "witch's brew" of chemicals. Even many plant-based dyes contain suspicious ingredients. Products aimed at the African American market, often containing straighteners, are even more suspiciously potent. Pregnant women are usually advised to stop dyeing their hair altogether because the dyes have a particularly pernicious impact on fetal development.

The wonder is that we still reach for those bottles, day after day, month after month, year after year. Ronnie explores the history of the beauty advertising industry that has risen up to convince us how deeply we need what's in those bottles. We are told—by a thriving cosmetic industry and the magazines and websites that are their willing, and financially dependent, colluders—that these chemicals color us in ways that are not only attractive but necessary. Necessary to maintain our sense of happiness, success, companionability, youth. Chemicals, regardless of their toxicity, are the key to being beautiful.

Needless to say, a book like this, stripping away, like peroxide, the layers of deceit around what our hair dyes are doing to us, makes me angry, very angry. And very alarmed. Although Ronnie is fastidious about not being judgmental about anyone's decision to dye her or his hair—after all, her story shows her own attachment to the evil genies in the bottles—about one-third of the way into reading it, I put down the manuscript and texted my younger sister to beg her to stop coloring her hair.

I feel anger at the FDA, for doing next to nothing to prove that our products are safe. I feel anger at the beauty industry, for playing fast and loose with information—and stoking our insecurities and preying on our vulnerabilities. I feel anger at the chemical industry, for using its vast resources to block reform. The only person I can't feel angry toward is my mother. For once.

I get it. We don't all have to love that salt-and-pepper look. But this book proves one important thing: we all have to rise up and demand, loudly, clearly, and passionately, that the FDA and the beauty industry do a better, more honest job of ensuring our safety. That the beauty industry stop selling products that contain chemicals that are linked to cancer, chemicals that disrupt our hormone systems. That it stop hiding behind the vagaries of just how difficult it is to pinpoint which chemicals cause which hideous disease over years and years of multiple exposures—just as the tobacco industry did so cynically, immorally. That they innovate toward safety.

We are racing for cures for heartbreaking diseases. We should also focus on another race: a race for the causes. Starting with those chemicals we slather on our bodies, and bake into our hair, day after day, month after month, year after year. We consumers are powerful—but not just because we can boycott products. That often results in what is called "regrettable substitutions" of one dangerous chemical for another.

We are powerful because we can raise our voices and demand that our lawmakers (who are, after all, mostly coloring their hair too, come to think of it) do better by us. That kind of strength doesn't come in a bottle. But it might just make what comes in those bottles safer, so that day after day, month after month, year after year, we are not poisoning ourselves.

That's real beauty.

Introduction

I'VE COME TO THINK OF this book as a negotiation between health and beauty, a conversation that started when the two collided with an obsession—hair. So, in the spirit of honest dialogue, I have to begin with two confessions: (1) I loved my dark dyed hair until it stopped loving me back, and (2) I'm not going to tell you to stop coloring.

Neither of these admissions sits easily with my role as an environmental health activist. For years, scientists have known that chemicals in personal care products were threatening women's health. Now, word of the questionable ingredients is out, not only on the street but down the aisles of Sephora. Even beauty magazines are publishing articles about phthalates, parabens, and other endocrine disruptors that interfere with the body's natural hormones.

Yet I spent twenty-five years coloring my tresses before I asked a colleague, an environmental scientist, "Are the chemicals in hair dye toxic?" Her answer was my wake-up call. It spurred me to begin asking a host of other questions and, ultimately, to write this book. What's the connection between coloring and cancer? Who regulates the hair dye industry? What are the risks for hairdressers? Are all dyes created equal? Do safer alternatives exist? Where does all that dye end up after it's washed out of our hair?

This truth-seeking investigation also led me to a more personal question. Given everything I was learning about the potential dangers of hair dye, why was the thought of giving it up so terrifying? I realized that for me, as for many women, hair dye was the "magic elixir" that made me feel youthful. Ditching the dye would mean confronting strongly held cultural beliefs—mine and others'—some of them so ingrained I was barely aware of them: beliefs about beauty, choice, aging, and femininity. It would also mean flouting fashion and beauty gurus, the media, decades of powerful and seductive advertising, my girlfriends, even the expectations of the men in my life.

Taking responsibility for our health is personal. My hair story is personal. How can I tell other women how to look or how to feel about gray hair? I can't. But I can give you two things I didn't have when, as a young mother, I began noticing my first gray hairs. First, information. Consumers deserve to know what's in the products we're massaging into our scalps. To make informed decisions about our health, we need to understand basic facts about questionable ingredients. And, with little transparency required from hair care companies, those facts are not always clear and not so easy to find.

Second, I can give you my own example and those of other women who are questioning the healthfulness of hair dye. When I started going gray, in my thirties, there were no silver-haired models gracing the covers of magazines. The thought of not coloring never even occurred to me. And when, twenty-five years later, I finally took the plunge and ditched the dye, I found myself facing tough beauty dilemmas. Would I opt for a short pixie cut? Blend my incoming gray with highlights? Go cold turkey and let the skunk stripe widen naturally? What about that infamous purple shampoo? And, ultimately, how old would it make me look?

One thing that made a big difference in reconciling my transition from dyed to natural hair was finding a community of silver sisters. From women leading by example, not only did I get some helpful tips (including eating oysters—who knew?), I also learned about a different kind of transition, the one that happens internally. While every woman has to find her own path, it helps to share the journey to self-acceptance—the grace of gray.

So, with a generous dose of humility, I've tried to create a guide to the complexities of environmental health and revelatory beauty for women of all ages. We all wonder if the choices we make will keep us healthy and whether they're worth it. Maybe you picked up this book because you've been toying with the what-ifs: what if I stop coloring and my hair looks awful, what if it makes me feel old, what if my boss or colleagues treat me differently, what if my partner doesn't find me sexy anymore, what if, what if . . .

Maybe you're not ready to give up coloring cold turkey. If that's the case, you'll find particular ingredients to avoid, safer alternatives, and a better understanding of the risks. Or, if you're contemplating a going gray transition, you can tuck away the information in this book and take it out again when you're ready to discover for yourself the sweet spot between hair, health, and beauty.

When we open ourselves to change, the hope—that most luxurious of feelings—is that we'll find our way to good health. It springs up because looking good matters only if you feel good. That's the prize—the rest is just hair.

Chapter 1

Hair Is Life

*H*ow'd you do it? *Are you doing that on purpose? Are you okay?* Ever since I stopped coloring my silver hair, I've gotten a lot of questions. One of the most common during my hair transition was *Why are you letting it go gray?* While my roots didn't ask permission before they stopped growing in dark brown, it was a complex mix of fear and determination that rearranged my beauty priorities. The question of *why*—why, after twenty-five years of using chemical dyes, I gave them up—is something I've thought about a lot.

My world began to shift four years ago. I was sitting in a meeting about toxics reform in Washington, DC, when an environmental scientist began to describe the buildup of chemicals in our bodies. As she rattled off a list of ingredients in personal care products—toluene, benzophenone, stearates, triclosan—my scalp started to tingle. "We're just beginning to understand how these chemicals compromise long-term health," she concluded.

None of this was new information. As a journalist, I report on the intersection of health and the environment.

I know that the soaps, shampoos, and lotions we use every day have been linked to threats such as hormone disruption, birth defects, and cancer. I know that since World War II, more than eighty thousand new chemicals have been invented. And while most people assume that the chemicals in our products have been tested and proven safe, I know that isn't the case. Time and time again, I've seen regulators fail to protect the health of citizens. Yet all that knowing didn't stop me from availing myself of the alchemical wonders of hair dye.

Frankly, coloring just seemed normal. My mother still dyed her hair a coppery brown at age eighty-eight, my best friend went to the colorist every few weeks, and even my daughter dabbled with highlights. It's no surprise that I didn't—and still don't—know many women who forgo coloring; 75 percent of women in the United States use hair dye. Like many, I colored with the hope of "natural-looking" hair, spending hours and hours, and thousands of dollars per year, at the salon.

Over the years, I'd pushed aside fears about the possible dangers of dye. After visits to the hairdresser, my scalp would itch, which I chalked up to dryness, and I would get headaches, which I blamed, like almost all other ailments during my childbearing years, on hormones. When I scratched my head, dye would stain my fingernails for days after application. A small price to pay for beauty, I rationalized. I simply did not want to think about the noxiously charged question *Is hair dye safe?*

A young colleague at the toxics meeting was more skeptical. Wiping the lipstick from her unlined lips, she asked, "Why do we subject our bodies to questionable chemicals?"

I could personally attest to the scientist's answer: "People ignore potential risks for convenience, cost, beauty. Many of these products promise a fountain of youth."

After the presentation, our group of mostly women discussed the health compromises people make "to look young and feel good." Scanning a handout with a long list of chemicals in personal care products, I decided it was past time to stop burying my beautifully dyed head in the sand. "How do I go about researching the toxicity of hair dye and its effects on me and others?" I asked.

The scientist's answer led me to the journey that would become this book. "The economic success of hair coloring collides so powerfully with popular demand that the task of understanding the landscape goes beyond science and law," she said. "Investigate that, and you'll find some answers that address the safety of hair dye."

~

In Japan, there's a saying, "A girl's hair is her life." It's a sign of female strength. Hair is a powerful expression of not only who we are but also who we aspire to be. It was against this backdrop that I came to love my long, thick dark hair. It was my most coveted beauty asset, a signature that told the world that I was unique and fun-loving and that I cared about a youthful appearance.

I owe that identity, in no small part, to my mother. Mom recalls having her hair braided by her own mother and living through World War II, when hairstyles were tailored and utilitarian. In her teens, she started to develop her own style. By the mid-1950s, when I made her a mother, *Vogue* had declared "hats and hair accessories as the must-have accoutrements of the day, while styling products hit the market."

She told me, "Hair is always important. It tells if you are well-kept and fashionable." I've been reminded of this my whole life, by her and by others. But my definition of "well-kept" differs wildly from my mom's. Ever since she had my long hair cut into a pixie when I was a child, to "make it easier to comb through," I started to create my own hair identity, which was (and still is) long.

By the time I was a teenager, I was doing what most teenage girls do—push their mother's buttons. I wanted to fit in with what was in vogue with my hippie-chic girlfriends. Finding a space to display teenage defiance, I pushed my hair obsession constantly.

It was the bane of Mom's existence. "Why must you bring your hairbrush to the dinner table?" she nagged.

Her admonishments came with a laundry list of dinner-time hair rules: no fidgeting with hair, brushes off the table, punctuated with "Hairstyling belongs in the bathroom."

Probably to annoy my mom, I didn't pay much attention to her Emily Post–like commands. This became an ongoing family joke between my brother and me, until my father lit up his cigarette before the rest of us reached for our dessert plates. Then Mom turned a disapproving gaze toward him for asphyxiating us with Chesterfield plumes. Horrified that he might billow smoke rings into my freshly washed tresses, I grabbed my brush, pushed my chair back from the table, and fled into my room.

To this day, I wonder why my mother imposed rules that only drove her into a tizzy. Did she think of the hairbrush as a dirty, germy object, while I equated brushing with cleanliness? Or was it just a maternal sense of control, just as her hair, a hair-sprayed helmet that towered to an unnatural height, was strangely untouchable to me?

Wanting to have nothing to do with hair that neither wind nor atomic bomb could penetrate, I brushed my smooth, polished straight hair incessantly to a shimmery shine that swung across my shoulders like Cher's. I also learned to keep my head far away from aerosols. Possible early nod to environmental health awareness? Doubtful. More like teenage rebellion against the sleek, hard crust of older women's hairstyles of that era.

I blame Veronica Lodge from the Archie Comics for sparking my hair obsession. Of course, she was also called Ronnie. With her blue-black sheen, my hair could have been drawn with the same thick charcoal pencil. The similarities between Veronica Lodge and the girl I used to be ended there. I did not grow up in the spoils of an affluent family. Poise and, luckily, a mean-girl attitude were not my thing. Nevertheless, by stature, we could have been synchronized swimmers in her parents' backyard pool. While I never sensed a whole lot of feminist thought going on in Veronica's self-possessed head, she was made of tough stuff. Coming of age in the '60s and '70s, with the civil rights movement, the Vietnam War, and the women's movement simmering to boiling points, even Veronica lived on the ledge of changing times.

She sets priorities straight in a comic strip with Archie and his dad:

"Say! How come you're sticking up for Betty?" Archie's dad asks, with Archie standing next to him, shaking in his boots.

"We're rivals only on trivial matters like boyfriends!" Veronica points at Archie with attitude. "When it comes to big issues like WOMEN'S RIGHTS we see eye-to-eye."

Like most teenage addictions of the time, mine was fed by the media. Cher and Veronica embodied not only sleek beauty but also self-assurance. For me, hair was power. But with age, my hair story became more complicated.

When I was a young mother in my late twenties and early thirties, gray hairs sprouted and quickly multiplied. To cover the gray strands, just a short time after my daughter, Lainey, was born, I began dabbling in hair color. When I look back at her birth photos, I can see flints of silver threads embroidering my bangs and part area. By the time my son, Ben, was born four years later, my hair was back to dark, dark brown.

Having children thrust me into a grown-up world. While I was figuring out my new role as a mom, I scrupulously reviewed every ingredient that passed my children's rosy lips. I learned about the campaign to ban Alar (daminozide), a chemical sprayed on apples so they would ripen before falling off the tree. Alar and the product it created when heated, unsymmetrical dimethylhydrazine (UDMH), were deemed "probable human carcinogens" after dangerous levels of pesticide residues were found in young children. When *60 Minutes* covered Alar in a segment titled "A Is for Apple," I switched to organic. Cancer risks from baby applesauce? Not on my mommy watch.

I dressed my children in natural fibers, hand knitting sweaters in cotton and wool. My farm-to-baby approach applied to the personal care products I used on their small bodies. After learning that about eighteen billion disposable diapers were thrown into landfills each year, and they take five hundred years to decompose, I switched from plastic to cloth.

Believe me, using cloth diapers was not easy. I knew only a handful of moms who would put up with the mess. But my children thrived, and that was what mattered most. And now, with everything we're learning about the disturbing list of fragrances, colored dyes, and chemicals in disposable diapers—including glyphosate, the notorious chemical in the herbicide Roundup—I'm glad I mostly used cloth diapers. The environmental impact of disposable diapers was an early wake-up call. I did not want to muck up our planet with plastic waste.

Ironically, this Earth Mama didn't think twice about covering her hair with chemicals that found their way not only into my body (and that of my nursing baby) but also into local waterways when the chemicals went spinning down the drain. This type of denial was a contradiction I wouldn't come to reconcile until years later.

Coloring didn't seem complicated in the beginning. One could choose a signature hue for life or be more exotic and go for a momentary beauty high. I chose life. Or so I thought. With unyielding certainty, I picked a color that was closest to my original dark, dark brown—because who wants to make that decision every few weeks?

At first, I took cover with a "semipermanent" or "temporary" darkest brown dye. I was told this would require a single-process procedure, just one application of color. In those days, my hair grew impossibly fast, and I washed it every day. I planned my life around those once-a-month three-hour salon appointments, getting babysitters and going as far as to make excuses to leave work early if it meant sticking to my coloring regime.

While hair coloring is a chemical process, a simple explanation is really all non-hairdressers need to know to understand how the different types of hair dye work to concoct their magic. Semipermanent color is temporary. It adds more tone and enhances natural color without changing hair color dramatically. In fact, it won't lighten hair because it contains no ammonia or peroxide. The dye fades after several shampoos, leaving the hair its original color. Hair care companies claim that semipermanent dyes can cover up to 50 percent of gray hair. To keep up this coloring routine, you need frequent applications, which, like all hair dyes, can cause damage.

I stayed a semipermanent gal for about ten years, and then three things happened:

1. Gray hair started to overtake the brown.
2. The semipermanent dye washed out to a faded muddy matte color.
3. The times between part touch-ups got shorter and shorter, making the dreaded skunk line appear sooner and sooner.

I was being drawn deeper into the hair dyeing cycle. Friends told me not to use permanent hair color unless I wanted to go lighter, which I did not. I had heard rumblings that permanent dark dyes were dangerous. I knew hair color used hydrogen peroxide, ammonia, chemical dyes, and long-lasting coloring agents. And with the combination of the stronger dyes in permanent hair color, the dye needed to be left on hair for longer periods of time, making permanent hair dye even more damaging. Despite the warnings and increased salon visits, I wanted to cover the gray. So I took the plunge and began using permanent hair dye for a "more significant color change."

In the first step of the process, ammonia and peroxide are used to lighten the natural pigment of hair and form a new base. Then the new color is added. The result is a combination of natural hair color and the chosen shade. I was surprised when my colorist told me she used the same permanent hair color on a friend of mine. My friend's color looked quite a bit lighter and redder. That was probably because her natural color was lighter and redder. With permanent hair dye, the chemicals need to stay on the hair longer. Regular touch-ups every four to six weeks are required to cover the regrowth at the root.

This Whack-a-Mole love affair with permanent color quickly came to dictate my schedule. Intensive monthly sessions eventually gave way to every-two-week salon visits as my skunk line moved in sooner and sooner. By the time I stopped coloring, I was going to the salon every other week to cover my part.

Even in the early days, the constant upkeep was a struggle. It came to a head one sunny afternoon as I sat in the bleachers at one of my son's Little League games, simmering with rage at an obnoxious dad who was yelling over and over, "Kill the pitcher!" My seven-year-old was the pitcher.

Fed up, I threw off my baseball cap and stood glaring in disbelief at the oaf. Teetering on the bleachers, I was about to launch into an "adults play fair" lecture when a friendly mom on the bench above me tapped me on the shoulder and said pleasantly, "Ronnie, are you doing that on purpose?"

Thinking she was also peeved about Little League injustices and hoping I had an ally who would help give the guy a lesson about acceptable behavior, I started to answer. "Yes, he needs to know how to act like a responsible . . ."

Peering down at the top of my head, she cheerfully said, "You're awfully young to go gray."

I was past due for a root touch-up, a social expiration date of sorts. I looked at my baseball hat lying in the mud five flights down. The hat was supposed to protect me from the sun and unsavory comments about my pariah part. But the hat had slipped from the floor of the bench in between the bleachers, and now I would need to climb down and crawl around in the damp grass to retrieve it. Lovely.

"Um, no. I'm not doing this on purpose," I stammered.

"Well, your silvers shimmer in the sun. Might look nice," she said warmly.

My son's honor was at stake, and all I could think of was, *Are you kidding? No freaking way.* I wasn't even forty yet. (I would later learn that approximately 32 percent of women are under the age of thirty when they notice their first gray hairs.)

Just then there was a burst of cheering from the parents on our team. My son had stopped walking miniature hitters and was on a spree of pitching perfect strikes. The crowd went wild. On my feet now, I glanced up to see the annoying dad watching me closely. He wasn't looking at my face to congratulate me. With a raised eyebrow and a cockeyed smirk, towering over me, he tipped his hat in the direction of the top of my head.

I lost the courage to take a verbal swing at the sweaty guy. *How could he possibly take me seriously with my unkempt part showing?* In that moment, I became the slovenly mom. Surrendering to an insane hair rule, I had only myself to blame.

Whoopi Goldberg once described how shocked she was when she first discovered that her "ass is bigger." She kept checking behind her and finally resolved that she was being stalked by her own behind.

Those slovenly roots stalked me everywhere, and they loomed large in my memory as, twenty years later, I made my way home to New York from the DC meeting on toxics. I was committed to researching dyes and their health effects. But could I give up coloring myself? Of course, the first person I consulted was my mother. My mom's hair is thinning, but her memory is thick with wisdom. Once I had blurted out my dilemma, she reminded me: "The world may be a different place from when I started coloring my hair, but old, graying hair is not necessarily good hair."

My mom had, and has, far more experience at the aging game than I have. Over the years, I've taken note of how she's dealt with physical changes—the lines and sags that settle on her face and body. She braces herself with the patience of a long-distance swimmer seeing the ocean for the first time. Taking a deep breath, she swan dives in and lets the consequences wash over her.

Since she has welcomed aging with compassionate grace, I wanted to find daughterly reassurance in her words. But I couldn't. It's hard to say exactly why I'm wired differently from my mom, but once I wade into deep waters, I find the current and fight it. I question everything, gnawing at answers until I'm satisfied. That's just how I bodysurf.

Bucking the norm raised all kinds of thorny beauty issues. I was way too vain to become a science experiment, so along with figuring out whether hair dye had, or would, hurt my health, I needed instructions on how to ditch it—how to go gray. How to go gray *and* protect my beauty. It was that old fear of being seen differently that tripped me up along the way to acknowledgment . . . and, eventually, acceptance. I was stuck floundering, a bombardment of doubts kicking in like a caffeine rush. The questions buzzed through my mind: Would I find a way to deal with the skunk roots without a color intervention . . . a fix . . . a straddling elixir that compelled me back to the sink with blackened water circling down the drain? Was there a less toxic hair dye? Has the wretched dark slurp already left its compensatory damage on my body, on my planet? And, with my misgivings reaching a crescendo, I asked myself, *How old will I look?*

To calm the war in my head, I banked on quickly finding answers to the practical question of how to transition from dye to gray hair. The existential questions were harder. I needed a comeback line to explain to myself and others why I was turning away from something that was such a beauty basic for women.

Ditching hair dye was a completely different advisory system from, say, changing facial moisturizers. Over the past few years, cosmetic companies have created healthier alternatives to slather over our faces, and now we have a few decent choices. But unless we're willing to go the route of cosmetic "enhancements," hair dye is our best bet for covering up age.

As a woman in my fifties, I had gotten the message loud and clear: hair dye is beauty pabulum. This was summed up beautifully in a beloved book, *I Feel Bad about My Neck*, by screenwriter and essayist Nora Ephron. "Hair dye has changed everything. . . . It's the most powerful weapon older women have against the youth culture . . . it actually succeeds at stopping the clock."

I worried my self-preservation would take a hit when I started to grow out my dyed hair. *I'm getting older, and now it's showing.* I would carry this burden around whether I wanted to or not. Facing my contradiction, I longed to lean into my age by going gray, but Nora's prevailing message still loomed large. My new "look" would certainly stand out in a jewel box of vibrant-toned hues that held very few gray gems.

In midlife, illness can be a possible reason for any change in appearance. I've seen that hair loss, as well as hair color changes, natural or not, can be an outward symptom of what's going on inside. Personal health becomes public when the knotty politics of aging and appearance come into play.

I didn't want people to think I was sick. And I certainly didn't want a different kind of pity: the kind of pity that befalls an actress who gives in to the ravages of aging, has missed a few Botox appointments, and finds herself on the unwanted receiving end of the cover of a supermarket tabloid. Busted by a beauty cop for giving up.

I found myself thinking of a *Seinfeld* episode in which Jerry confronts George about wearing baggy sweatpants: "You know the message you're sending out to the world with these sweatpants? You're telling the world, 'I give up. I can't compete in normal society. I'm miserable, so I might as well be comfortable.'"

She's letting herself go. Let's be clear: as we move through life, we don't "let" our hair go gray. It *grows* gray. We cover it up. I didn't want to be pinned down by an aging beauty standard about gray hair, and that led to an intense need to be free from the trappings of hair coloring. Becoming increasingly aware of the effect chemicals from personal care products have on women's bodies, I decided to stop making bargains with my health and became open to the possibility of beauty after coloring. It didn't come easy, and I know, and respect, that it's not the same for everyone.

With mixed-up emotions about age, vanity, and health, I decided to look to three important women in my life to find out how hair shaped their world.

"Hair is a defining trait, but it does not grow in my favor," my best friend laments. I've known Cathy since junior high school, and I know that, despite her petite stature and athletic grace, her sense of beauty has always been wrapped up in what she considers a "bad hair package." She believes the fine texture of her dyed brown hair has been handed down through generations. In the past, the women in her family covered their thin hair. Cathy once confessed that her grandmother had a whole room devoted to wigs. In Cathy's estimation, the thin hair gene hit the men in her family too. "My brother started going bald at twenty. So there was no hope for me." Would she ever stop coloring? "If I had thick hair, maybe I would consider it."

My sister-in-law, Carol, is a gorgeous Latina who has balanced beauty, brains, and brawn throughout her life. As a child, she wasn't given the choice to cut her hair because her mother insisted, "It's traditional for Mexican women to have long hair." When she left home for college and became a Stanford University cheerleader and a model, her exceptionally long, shiny black hair made her feel exotic, but it was difficult to manage. I met her when she was twenty-two and almost exclusively wore her hair in braids. Carol longed to cut her hair short, which she finally did when she became a doctor and "started a new life." A two-time survivor of both breast and uterine cancer, Carol says that when her sister went prematurely gray in high school, her mother permitted early hair coloring. When Carol started to go gray, just a few years ago in her early fifties, her mother took one look at her and announced that she should immediately color. Carol tells me she loves the way the silver strands highlight her still mostly black hair. She says that instead of coloring every four to six weeks, she treats herself to a sexy, chic short haircut.

Cool, hip, and stylish, my daughter, Lainey, possesses a striking beauty. Yet she sees her wavy brown hair as an "out-of-control mess." Aside from a brief stint of straight hair during puberty—burgeoning hormones?—she has struggled with her curls. She tells me that growing up, "I had to accept the fact that my hair wasn't the slick, straight look of yours, but it also wasn't the beautiful perfect waves I also longed for." In high school and beyond, she straightened her hair with a flat iron because she never felt "put together" with curly hair. When she started flirting with blond high-lights in high school, she developed a newfound love for her hair. "I was going for the beachy/surfer/Cali girl look, even though I lived in upstate New York. The highlights gave me confidence." She claims to be a little more at peace with the texture of her hair, but now, in her early thirties, she's not thrilled about getting gray. "I find myself back in that self-conscious teenage mind-set. I want to cover up! And, as you write a book about how questionable hair color might be for my health, I find myself feeling guilty and torn. I'm not ready to succumb to the older white-haired look, so I feel a superficial weight hanging over my head (no pun intended) for wanting to keep my 'natural' color, which would mean dyeing it. Knowing what I know, and not wanting my hair to look gray, what do I do?"

Pleasure, pain, control, sacrifice, guilt, good and bad days crackle through the tangled strands of women's hair stories. Despite being encouraged to discipline our hair to conform to an ideal, we take stock of our unique beauty and express ourselves individually.

With that resolve, I finally figured out a way to answer my friends' well-meaning questions, a shorthand that explained both the practical and fundamental reasons I was ditching the dye: "The upkeep, the cost, the chemicals."

Armed with my mantra, I peered in the mirror, at the roots that would soon shine back their moonglow. I could only guess that once I abandoned my biweekly root coloring, I would have a glittering crown of silver hair.

Remembering that appearance matters if our bodies stay healthy, I confronted an uncomfortable reality: hair color had become a beauty life raft. I asked myself, with a long, healthy life as my end goal, what was I willing to risk to look younger? So, without surrendering my vanity, I set the hair color myth adrift (sorry, Nora), sensing it really wouldn't save me from aging, hoping to discover a brave new "hair is life" story.

Chapter 2

Hair-Raising Salons

I T'S NOT HAIRCUTS, BLOWOUTS, OR STYLING that drives the salon business. It's professional coloring. In fact, the industry refers to hair color as "the vital anchor service" that draws customers to salons across the United States. Once there, women often spring for cuts, perms, straightening, and other treatments. But color is the core mainstay. And, according to a 2018 *InStyle* survey of almost 1,500 women, hair color services are up; only 7 percent of the respondents were dye free. The reason for the upswing? Baby boomers in need of covering gray hair.

Hairdressers play a unique role in the salon business: part hair worker, part therapist, and part guardian of secrets hidden under colored locks. *Does she or doesn't she?* As with other beauty services, the relationships between customers and hairdressers are built on a delicate web of loyalty and trust. Most women pick salons and stylists very carefully, since this intimate bond can last for years.

At the time I decided to stop coloring, my go-to gal for all things hair was Heather. I inherited Heather from her lovely mild-mannered aunt and ex-military husband

21

turned mountain man. This breezy hairdressing duo had tenderly clipped my babies' soft curls as they squirmed in my lap, tamed my husband's ultracurly locks, and set me on a colorful path for covering my grays. In their haircutting heyday, we were entwined in each other's lives, exchanging support throughout almost two decades.

What attracted me to this hairdressing clan was their sincere passion for all things "natural." In fact, they named their salon Supernatural before it was even a riff on Carlos Santana's guitar. At the time, Supernatural seemed a beacon among the local salons because it was one of the first of its kind to dish out Aveda hair color. Lured by promises of "96 percent naturally derived, essentially damage-free" color, I found my groove with Aveda color and products.

Now, having finally figured out that "natural permanent color" was an oxymoron, I knew I was looking at a two- to three-year hair transformation. I needed Heather to throw out her standard rulebook, develop a *transition* plan, and help me get to gray. But I worried that my new approach to natural beauty would test our relationship.

I sure hoped I wouldn't encounter something like the sparks that recently flew when the United Kingdom's leading celebrity hairdresser, Nicky Clarke, shamed the Duchess of Cambridge for stepping out with a couple of inches of gray roots catching the sunlight. Opining in the pages of the *Daily Mail*, Nicky wrote, "Until you're really old, you can't be seen to have any grey hairs." Except if you're a man. "Men can go grey in their mid-50s and still be considered attractive," Nicky claimed, citing the whole "Silver Fox thing." But Nicky didn't see it the same way for women. A few strands of gray would surely knock the duchess off her style icon pedestal. He concluded, "I highly recommend that she cover it up. I hate grey hair." Ouch.

Sarah Harris, a silver-haired editor for *British Vogue*, snapped back, "To cast such trite aspersions is like saying that women can't have long hair the other side of 40." Sarah countered Nicky's low blow, referring to him as a fifty-seven-year-old man who shouldn't be allowed to have "a blond, flowing, tonged (?), highlighted (?), backcombed (!) bouffant, whether they're a celebrity hairdresser or otherwise." Double ouch.

Wanting to keep that kind of flap on the other side of the pond, I braced myself as I stepped into the salon. I was instantly surrounded by the familiar buzz of hair dryers, the whiff of chemicals, and the sight of those creepy circle swatches of artificial hair in various shades. All of it now seemed like the choreography of an old, out-of-step dance.

In the mirror behind the half wall, I could see Heather in cape and gloves, slapping her magic wand into the brackish glop in preparation for my root covering. With her amazing chop shop skills, long bluish-black hair, and permanently stained fingers, multi-talented Heather had endured women's hair obsessions since she was a teenager. What I loved most was how she took her time and executed my wishes with razor-like precision.

"Just a trim today; I'd like to stop coloring." I held my breath as Heather intensely examined my roots. Those roots. The ones that would start growing the minute I walked out the door.

"How and why would you like to do that?" she asked sweetly, taking her gloves off and staring down at her raw hands as though I hadn't just thrown a Molotov cocktail into the world of bottled hair color.

Looking around for answers, I noticed a youthquake of models in silky, sexy hair posters on the walls, enticing me to embrace "vibrant, fade-resistant color with amazing shine." I thought of legendary fashion designer Miuccia Prada's comment that women try to tame themselves when they get older, but instead they should strive to be wilder.

Where were the fashionable older women who had recently graced the Style section of the *New York Times*? The interesting ones described as "women who have fun. [Gray hair] reflects their confidence, their ease with being who they are."

With resistance mounting, it dawned on me that I had been reading those posters for years, and yet I had never asked to examine the ingredients in my own hair dye.

Beauty and Its Beast

"Follow the chemicals, and then decide if it's worth the gamble," advised my father-in-law, a retired chemist, when I told him I was thinking about shutting down the color.

His words replayed in my head as I read the chemical names on the safety data sheet of my hair color. In retrospect, I really wasn't sure how I had managed this level of denial—me, the environmental activist who reads every label like an FBI agent reopening a cold case, poring over it for new clues. Why hadn't I asked to read the ingredient list sooner?

. . . *phenylenediaminepersulfateshydrogenperoxidelead-acetatetoluene* . . .

The words on the sheet bled together, tumbling into one chaotic blur. In my tunnel vision, I could barely hear Heather's transition plan. As she combed through my dark strands, uncovering a one-inch seedbed of gray, I worried about hair dye's effect not only on my own body but also on hers. Styling hair is the equivalent of working in a chemistry lab. I couldn't help thinking about the old adage "The dose makes the poison." Heather inhaled these chemicals every day, they were absorbed by her skin, yet she didn't recognize the risk.

But I knew. I knew because as editorial director of Moms Clean Air Force (MCAF), a million-plus member special project of the Environmental Defense Fund (EDF), I received a daily stream of aggregated news about toxic chemicals.

I've learned that in the United States, over 85,000 chemicals make their way into our bodies. They come from contaminated water, pesticides in food, air pollution, household cleaners, and personal care products. While Canada and the European Union have banned over 1,300 ingredients from use in cosmetics (including some hair dyes), the United States has banned only 11.

From mercury in mascara to styrene in maxi pads, toxic chemicals in products expose women to over 100 chemicals each day through personal care products. Those small daily exposures can lead to chemical buildup in our bodies.

When I made my decision to stop coloring, I printed out a groundbreaking 2014 report from Women's Voices for the Earth (WVE) titled "Beauty and Its Beast," and tucked it into a file. It highlights decades of research on hair salons, including the harmful chemicals in hair products and how they impact a predominately female workforce.

A toxic brew of chemicals is commonly found in the air of indoor salons, which generally have little or no ventilation. This shouldn't be surprising, given the ingredients in hair products, including formaldehyde (known carcinogen), toluene (neurological and developmental toxicant), sodium hydroxide (lye), and triphenyl phosphate (suspected endocrine disruptor).

Surveys have found that 60 percent of salon workers who work with hair dyes, bleaches, and permanent wave solutions suffer from skin conditions, such as dermatitis on their hands, starting as far back as cosmetology school training. *Heather's stained hands.* Salon workers also tend to have higher rates of miscarriages, gestational diabetes, and babies with birth defects. And they have a greater risk of dying from neurological conditions including Alzheimer's disease, presenile dementia, and motor neuron disease.

Each threat to hairdressers' health is serious, but one stands out from the rest: cancer. The International Agency for Research on Cancer (IARC) has found that "occupational exposures as a hairdresser or barber are probably carcinogenic to humans." Other studies have found that hairdressers were three times more likely to get breast cancer than women in other occupations and five times more likely to get bladder cancer than the general population. And that risk is even higher for black women. Other increased cancer risks included lung cancer, laryngeal cancer, and multiple myeloma.

The Shade Makes the Poison

While Heather faces increased risks from exposure to dyes day in and day out, the statistics for customers like me are hardly more encouraging. For instance, a University of Southern California study found that women who had colored their hair once per month for fifteen years or more had a 50 percent higher risk of bladder cancer. And it turns out that the color you choose plays a role in the level of danger the dye poses.

The deep-brown shade that gave me such pride as a teenager can be particularly harmful when it comes from a dye bottle. Many dark dyes, sometimes called coal-tar dyes, include para-phenylenediamine (PPD), a chemical substance derived from petroleum. These dyes date back to the turn of the twentieth century. In the late 1800s, it was discovered that hydrogen peroxide could be used as a bleaching agent, paving the way for women (and a few men) to enter the world of commercial hair color. French chemist Eugène Schueller based his 1907 hair color invention on PPD, which, when oxidized, turned the hair black. Ironically, the name of Schueller's company was the French Harmless Hair Dye Company, later changed to L'Oréal. The first company to use PPDs in the United States was Clairol.

Today, not much has changed. PPD, which is also used in antifreeze, is still found in most permanent hair color; a similar compound, para-toluenediamine (PTD), is sometimes added instead. The Environmental Working Group (EWG) Skin Deep website, a database of ingredients in more than 41,000 personal care products, rates PPD a seven out of ten in terms of toxicity.

Dark dyes, along with chemical relaxers, have put African American women at increased risk. According to a 2017 study of more than 4,000 women, conducted by researchers at Rutgers University, use of dark dyes by black women was associated with a 51 percent increased overall risk of breast cancer—and a 72 percent increased risk of estrogen receptor positive breast cancer. The Centers for Disease Control and Prevention (CDC) adds that African American women are 34 percent more likely than white women to die of breast cancer.

In addition to increasing cancer risk, PPD can trigger itchiness, redness, irritation, and even fatal anaphylactic reactions. And even if you've been using the same hair color with no ill effects for years, or, conversely, if it's the first time you've ever tried hair color, the reaction can be severe.

In my early forties, a few years after I started dyeing, I developed unexplained allergies. My symptoms ran the gamut from hives to rashes to exercise-induced asthma to dangerous anaphylactic reactions. Connected? When I asked the allergist, he shrugged his shoulders and said, "You're a healthy girl with a bad problem." We were never able to get to the root of it. Now I carry an EpiPen with me at all times, and I've suspected my intense coloring routine may have played a role in triggering my allergies.

During the twentieth century, allergic reactions to PPD became so common that its use in hair dyes was banned in Germany, France, and Sweden. Yet it is still used in 80 to 90 percent of hair dyes in US salons and drugstores. In an article in *US News and World Report* titled "Are You Allergic to Hair Dye?" Dr. Andrew Scheman, an associate professor of clinical dermatology at Northwestern University, is quoted as saying that even most so-called natural hair dyes on the market are really just gimmicks. "They're not natural at all," he says. Like other hair dyes, he explains, many brands that claim to be natural contain PPD with a few extra botanical ingredients thrown in.

Bob Hefford, a former chemist for Clairol and Unilever, admits, "It is most probably true that if these materials [PPD and its cousin PTD] were invented today, their use in cosmetics would not be permitted but they remain in use as no effective replacements have been found."

If certain chemicals present a clear danger to the health of hair salon workers and their clients, particularly those using dark dyes, it seems we could simply avoid using them. Unfortunately, it's not that straightforward. While manufacturers of consumer cosmetics are legally required to print their ingredients on the label, makers of professional cosmetics do not have to disclose their ingredients. Instead, they're required to produce safety data sheets, which may or may not present the whole story.

According to the "Beauty and Its Beast" report, some safety data sheets are comprehensive, but those are "few and far between." And there is little enforcement to ensure the sheets have all the health information that hairstylists and customers need. I noticed that the print on my hair dye safety data sheet was so tiny that it was almost impossible to read without a magnifying glass, even with my glasses on.

Without the advantage of industry transparency, neither hairdressers nor their clients know what they are being exposed to, even as their bodies may be exhibiting an array of documented maladies. While the details may be hidden from women, many are now catching on to the threat. Researchers at Northwestern's Feinberg School of Medicine analyzed complaints made to the Food and Drug Administration (FDA) between 2004 and 2016 about adverse reactions to products and found that hair care products topped the list for women. Reports of serious health issues were also significantly higher than average for hair care and hair color products.

Pushed Out of the Nest

With the alarming "Beauty and Its Beast" report fresh in my mind, I was finally feeling like I could rise above this toxic cloud. I was committed to my decision and began to see the silver lining. But before I had a chance to tell Heather about my light bulb moment, my new hair mantra—*the upkeep, the cost, the chemicals*, and some of these other hair-raising tidbits, she unfurled a challenge.

"You have two choices. Lowlights or chop? Ronnie . . . ," she said with an intimacy that brought on goosebumps, "you're all about your long hair, and growing in gray hair will wash out your complexion. So I suggest we give you a multi-tonal effect, a few lowlights and a shorter cut. You don't want to look like you've given up, do you?"

Oh boy, here we go again. Each time someone mentions "giving up," an unmistakable queasiness takes over, threatening to flatten me. Heather held an outsize role in my beauty, in my life. Forcing myself to listen to her "plan," I could tell she was trying to convince me that one false step could send me down an endless spiral of bad hair days. All of which were a warning sure to age me light-years.

I worried about knocking her off balance. Since we've had such a close bond, Heather's told me about how hard her life has been as a single parent, trying to keep her young daughter fed, clothed, and sheltered. In that moment, I envisioned the little phoebe bird that each year builds its nest in one of the light fixtures outside our front door. I marvel that this small creature chooses such a busy location; each time the screen door slams, she flits from her nest. When the summer sky opens to a deluge of windswept water, threatening to blow the little nest to smithereens, Phoebe sticks her chest out and stands strong. Despite all the disturbances, she keeps showing up year after year to tend to her latest brood. Their wide-open beaks and tiny fluffy bodies depend solely on her, their meal ticket until they take flight to fend for themselves.

Afraid of losing my nerve, I blurted, perhaps a little too loudly, "I'll just let it grow out! Natural. No color." Even over the drone of blow-dryers, heads turned.

To soften the blow, I decided to relinquish some control. As a consolation prize, I let Heather apply copious amounts of products (no dye) before she ran over to the appointment desk to scratch out a year's worth of biweekly touch-ups.

Perplexed, the velvety-haired receptionist asked, "Why? Ronnie loved her hair color. It made her look so much younger."

"Her hair stopped loving her back," I heard Heather whisper. She shook her head as she stared down at the appointment book.

As I left the hustle and bustle of the salon, traipsing down the long flight of stairs that led out to the street, I heard the receptionist's voice trail off. "Did you remind her of the cancellation policy?"

I should have known that asking a hair colorist, particularly one who depends on a revolving clientele, to stop coloring my hair would be like asking a candy shop owner if you should give up chocolate.

Leaving with just a sticky trim, I realized that my journey from my high-maintenance color to "natural" (whatever that was) had begun. Without the security of the salon, it would be only a matter of days before a skunk line took up residence on my part. It was scary but also intensely freeing. I repeated my mantra—*the upkeep, the cost, the chemicals*—willing the dye to loosen its magic grip on my psyche. Thus I ended my long love affair with hair color.

And so, during this season of uncoloring, in the run up to sixty, I did not set foot in a salon except for the occasional blowout. Instead, armed with a kaleidoscope of questions, I focused on the safety of hair dye, hoping my newfound knowledge would help pull me through the long "how to do it" phase. As luck would have it, gray hair seemed to be on the cusp of a transformational cultural moment for women. The next step was practical. I needed to find a new hairdresser willing to embrace what I knew in my heart would lead to a healthier hair story. I needed one ASAP.

I've Got Just the Guy for You

In Diane Keaton's autobiography, *Let's Just Say It Wasn't Pretty*, she spills the beans on a few of her former lovers. In a passage about Warren Beatty, who played the sexy hairdresser in the 1975 film *Shampoo*, she says, "Warren used to pontificate on the subject for hours, insisting that hairdressers were worth their weight in gold. According to him, hair was, in fact, 60 percent of good looks."

I required a passionate hairdresser (hold the sex) who understood that in the golden land of women's beauty, hair occupies the prime location. And I needed one who respected women who chose to go natural. I found him amid the squash blossoms and shishito peppers at my local farmer's market. Richard, with his longish black hair and dark-framed glasses, came to my rescue via an acquaintance. Noticing that my graying hair was in need of pruning, she exclaimed, in the middle of the market, at the top of her lungs, "I've got just the guy for you." Living in a small town has its interesting moments. This was one of them as all eyes focused on me and the wheel of the gossip mill began to spin.

Richard became the epicenter of my hair transition. He had incredible styling cred, including stints at Oribe and Sally Hershberger, and was then working in New York City's Bumble and Bumble salon. Plus, he knew how to soothe women's hair obsessions. But that's not the very best thing about Richard. Amazingly, Richard made house calls! His best friend lived near my town in the Hudson Valley of New York, and Richard visited every month or so to grab a beer with his friend and cut a few "freelance" clients. When he came to my home in the woods every five to six weeks to chop away at the dead dyed ends, I would drag out a private Pinterest "Silver" inspiration board loaded up with photos of chic gray haircuts. Sans judgment, Richard would examine my favorite photos and cheerfully exclaim, "That would work!" When it comes to silver hair, he's a fan of any length, "as long as it doesn't look witchy." He assured me, "You'll be able to carry off most lengths during your transition. So you'll have no need to pull out the broomstick."

In front of the mirror, in the privacy of my bathroom, as my hair ever so slowly transitioned from darkest brown to silver, Richard and I discussed much: his younger girlfriend, whom he'd like to marry; child-rearing through the ages (he's a divorced dad of a school-age girl—I'm an empty nester with two twenty-somethings); meeting his idol, Bruce Springsteen, at a runway show (Richard's a hardcore New Jersey guy); Meg Ryan's choppy short hair. Who knew it was so thick? How does he know? "Dish," I plead. When I test him tepidly, asking whether I should get the show on the road and cut really short, he knows I can't. Or maybe he just doesn't want to see me cry? Richard has become my Warren Beatty—*worth his weight in gold*—minus the tight pants and philandering.

Split Ends

One day when I was sitting on a stool in front of the mirror while Richard piloted his scissors, elbowing around my small bathroom, he mentioned that he hadn't been feeling well. He had developed asthma a while ago. He was quitting his latest job at a salon in Brooklyn and moving upstate to be closer to his daughter.

Cut off from the drama of the salon scene, Richard and I discussed my latest hair obsession: toxic chemicals in the salon biz. By that point, the scary statistics had become firmly wedged into my consciousness, and I couldn't help but make the connection between hair treatments and his ailment. He was unaware that the keratin straightening treatment that he'd been applying to women's hair contains formaldehyde. The chemical is a potent allergen that is severely irritating to the eyes, nose, lungs, and throat, and long-term exposure has been linked to an increased risk of cancer.

My best friend, Cathy, had been toying on and off for years with trying out professional straightening products to transform her curly hair. At the time, I did some preliminary research and learned that smooth, frizz-free hair can last up to twelve weeks, but it comes with a risk. The most popular brand, Brazilian Blowout, contains formaldehyde, which is released as the hair is heated with dryers or curling irons.

I emailed both Cathy and Richard the "Beauty and Its Beast" report fact sheet. It specifically states that hairdressers have measured decreased lung function and higher risks of developing asthma. Cathy, a nurse, changed her mind and decided to forgo the risk. Richard promised to share the fact sheet with his colleagues because the list of other illnesses is also frightening: miscarriages, neurological disorders, immune disorders, dermatitis, depression, and cancer.

"How can 'they' let this health threat happen?" Richard asked incredulously. I told him the simple answer: "Because it's not a regulated industry, so anything goes. Someone has to get sick before the FDA can do anything."

Women's Voices for the Earth (WVE) and the Environmental Working Group (EWG) have been working to get the FDA to join California, Oregon, Canada, France, and Ireland in taking action against hair products such as Brazilian Blowout. They would like to legislate the removal of products with dangerous levels of formaldehyde.

"Salon workers have particularly suffered due to symptoms associated with these products, with many reporting long-term health problems," said Alexandra Scranton, director of science and research for WVE. When the health risks of using Brazilian Blowout were brought to the FDA's attention in 2008, the agency did nothing. Scranton finds this unconscionable.

Kathy Langford, a professional hairstylist from Brandon, Florida, points out that both hairdressers and clients are "exposed to the noxious fumes when these products are heated." And Kelly Merriman, a stylist from Joliet, Illinois, worries that when her clients return home after a treatment, they will unknowingly expose their children to formaldehyde.

If this isn't terrifying enough, these straightening products are labeled for professional use only. As we now know, regulation of professional products is stunningly lax, as companies are not required to list their ingredients on those safety data sheets. In the case of Brazilian Blowout, the Occupational Safety and Health Administration (OSHA) is investigating questions and complaints from hair salon owners and workers about possible formaldehyde exposure. And the FDA issued a warning letter to the importer and distributor. The letter identified the product as adulterated and misbranded because it contains methylene glycol, which can release formaldehyde during normal conditions of use, and because the label makes misleading statements ("Formaldehyde Free" or "No Formaldehyde").

Once a person becomes sensitive to formaldehyde, even a low-level exposure to the chemical can cause the body to react, and the sensitization may not be reversible. Had Richard become sensitized to those chemicals? Certainly the methylene glycol solution that he inhaled may have caused the onset of his asthma.

After a period of being voluntarily out of work and freelancing, Richard has decided to follow his dream and open a barbershop that provides no chemical processes—no Brazilian Blowouts, no hair extensions, and no hair dye.

"I'm going to call the new shop Barber and Brew," he tells me. "Men and women can come in for a cut and a beer. A simple, slow hair approach with a happy ending—a local handcrafted beer."

It sounds dreamy, and I encourage him to go for it. But I wonder why he won't provide hair extensions. Then I realize I know almost nothing about them, except that I have a friend who miraculously went from shoulder-length to down-to-her-waist-length hair. I also remember reading in a tabloid a few years ago (at the salon!) that the woman who looks like she's never had a bad hair day in her life, Jennifer Aniston—a certified hair icon—had chopped off her long-layered hair for an angled bob because her real hair was getting thinned out and damaged by hair extensions. Who knew that wasn't all her gorgeous hair?

I ask Richard why he won't apply hair extensions. He shakes his head and grimaces. "They can smell bad."

While there's a laundry list of toxic substances spewing from hair salons, glue wasn't one I had considered. But it is one of the ways the extensions are attached to natural hair. (Extensions can also be keratin bonded, sewn or woven in, or clipped or taped on.)

"Some glues contain styrene, which is a carcinogen," says Jamie McConnell, director of programs and policy for WVE. She also mentions other chemicals, trichloroethylene and dioxanes, that can damage the liver and kidneys, but she qualifies this by saying that not all extensions contain them. You need to be savvy and ask for the safety data sheets. If you can't find one, consider that a red flag.

With warning flags flying high all over salons, how is exposure to multiple chemicals affecting stylists and customers? Dr. Bruce Lanphear, an environmental health expert, says we need to take seriously the question of whether or not safe levels exist. Even low levels of some chemicals appear to be "proportionally more harmful to a person's health." In a new study reported in *Environmental Health News*, not only are there no apparent safe levels or thresholds of some of the most common, extensively tested chemicals, such as radon, lead, particulate matter, asbestos, tobacco, and benzene—but also at the lowest levels of exposure, there is a steeper increase of risk. Dr. Lanphear argues that most health and regulatory agencies are not fully protecting public health because they don't target people who have low to moderate exposures.

So maybe the dose does not always make the poison? Leah Segedie, author of *Green Enough*, explains: "In recent years, more and more cases have been discovered—particularly with endocrine-disrupting chemicals (EDCs)—where the effects do not follow this pattern. In some cases, the lowest dose tested has huge adverse effects, whereas a much higher dose has no impact at all." She goes on to note that EDCs have effects at low doses that are not predicted by effects at higher doses. And this is important because those who test and regulate chemicals for safety apply the adage "the dose makes the poison" with the assumption a low dose is less harmful.

If manufacturers are not required to provide their products' safety information to the FDA, that leaves salons, hairdressers, and consumers to vet and fend for themselves. Then they must evaluate whether or not the risk is worth it.

Right now, the chemicals in hair dye, straightening products, and other hair products carry both known and unknown potential health hazards. But the tea leaves show a tale of pollution—body pollution that could lead to disease. So, is it impossible to believe that quitting hair dye will make me less vulnerable? I'll have to live with that question to find out.

In the meantime, with Santana's "Smooth" humming around my head, I've doubled down my resolve to stay super natural . . . "You got the kind of lovin' that can be so smooth. Give me your heart, make it real, or else forget about it."

Chapter 3

The Beauty of Authenticity

I HAD JUST BOARDED a New York City–bound train in Washington, DC, when a guy with an Amtrak patch on his coveralls yelled out: "Hey lady, nice hair. Bet you didn't find that color in a bottle!"

In the midst of the travel blur, commuters darting onto the train paused to look. I kept my head down and settled into an available seat next to a businessman. Slightly embarrassed by the attention from the train worker, I focused my gaze out the window, trying to act aloof as the track patterns came into view. "You do have good hair," the suited guy said, nodding his head in my direction. He was around my age, which in my magical thinking placed him anywhere between fifty and seventy.

Ugh, hair commotion. Somewhere during the two-toned stage, before full transition, I stopped being hyper-aware of this odd grow-out phase. Yet I still thought of myself as having "good" hair. Just not good transitioning to *gray* hair.

Mildly flattered, I took the bait. "Do you always comment on women's hair, or is it just my crazy two-toned mop?"

"Your hair is refreshing, natural. You look comfortable in your skin—at home in your world." Humming a Billy Joel song, he chirped, "Only the young dye good?"

His twist on the Long Island crooner's song I grew up with made me smile. When he asked me about my work, he commented that it was "good work." We went on to enjoy a lively conversation about what middle-aged parents talk about: their adult children.

We exited the train, and as I was getting my bearings in Penn Station to locate the track heading to Rhinecliff, he thanked me for looking honest. Then the man dug into his briefcase, handed me his card, and said, "If you ever need a dermatologist who won't promise anti-aging, please give me a call."

I asked, "What do you mean, you don't promise anti-aging?"

He paused a minute and replied, "At our age, why would we be anti-aging?"

Exactly, I thought. After all, what's the alternative? Although I was a bit surprised that a dermatologist would admit that we can't look young forever.

Recently, *Allure* magazine announced it would retire the term "anti-aging." A bold move, considering that Americans are projected to spend almost $350 billion by the end of the current year on anti-aging products. The largest buyers of anti-aging products, not surprisingly, are . . . you guessed it: older women. Instead of anti-aging, the beauty magazine's editor in chief, Michelle Lee, wanted to encourage a celebration of "growing into your own skin—wrinkles and all." Lee explained that the term reinforces a messy notion that aging is something we need to battle.

It's no secret we are a youth-obsessed culture. Advancements in health and beauty may have given us more choices to control our looks, but the pressure to hold off the appearance of aging continues to send shudders through most women. This is why I found actress Amanda Peet's candid *Lenny Letter* essay, "Never Crossing the Botox Rubicon," refreshing. In the viral essay, the actress draws a line in the sand, giving herself over to the telltale signs of aging without cosmetic surgery.

Botox may be a rarity where you and I live—and, of course, we may never know who's gone under the knife and who hasn't—but in the celebrity world, cosmetic surgery reigns supreme. It's the norm. And let's not forget that cosmetic surgery is also called plastic surgery. I won't go down the rabbit hole of plastic pollution and body pollution; the name says it all.

With social media images readily available 24-7, women in and outside the entertainment industry alike are bombarded with unrealistic beauty ideals. And for some, like Peet, those ideals control job prospects. The forty-four-year-old mother writes about losing an acting gig to a seamless twenty-seven-year-old because Peet wasn't "current enough." She wavers in this valley of doubt: "The train has left the station and I'm one of those moronic stragglers running alongside with her purse caught in the door. Everyone's looking at me like, 'Let go, you bullheaded old hag! There's no room for you.'"

Worrying about the message this sends her young daughters, Peet retreats from even considering altering her face with surgery. She sees her girls learning that her "employability is based on looks," and she imagines them someday accusing her of being "nothing but a foot soldier for the beauty industrial complex."

Those beautiful daughters Peet is raising to embrace a healthy body image will one day look deep into their mother's face and see their story. Who would ever want to erase that?

There is no question that aging both gives us the experiences that form our lives and makes us more vulnerable. We lose not just our hair color but also our fertility, often our flexibility and strength, even our damn eyesight. Yet fighting aging as if it were a disease is a losing battle. You can't defeat time; you can only decide how to spend it. I refused to waste any more of mine on trying to appear younger.

Yes, I wanted to look good and feel good, but I also wanted to do good. Like Peet and many other women, I make choices with my children in mind. We have a campaign at Moms Clean Air Force called "Baby Power: Every Baby Has the Power to Make Us Change the World!" The initiative draws on a familiar idea: that babies bring out new virtues in their parents, particularly their mothers. I know that when my children, Lainey and Ben, were born, something flipped in my psyche and I became consumed with a newfound drive for protection. The world felt scarier, more dangerous. Wallowing in fear wasn't an option—not when I needed to be present and compassionate—so I threw myself into making my children's environment safer. Yet no matter how hard we strive to protect our kids, we mothers question ourselves: "Am I mom enough?" I'm discovering it takes a lifetime of parenting to find the answer.

Basic maternal instinct is one of the reasons I'm fascinated with environmental health. It's a passion I share with my husband, who also works in the environmental field. I've learned how environmental threats and inadequate or nonexistent government regulations can harm the most vulnerable: children. My kids inherited my concern, making healthy choices and sharing that lifestyle with their partners.

By shutting down my use of hair dye, I felt I was setting a good example for my daughter. Yet when she gave me the happy news that she was getting married, I hit a temporary derailment. After the initial blush of excitement, I thought, "Oh no, my hair is not fully transitioned yet!"

It was as if all the hard work I had done, both physically (stopping the dye cold turkey) and emotionally (shifting those pesky anti-aging thoughts), were suddenly being called into question. On one hand, I relished this celebratory time. I was so thrilled that my wonderful daughter had found the love of her life. On the other hand, I wasn't proud of my feelings about myself. For the wedding, I wanted to look my best, and I started to yearn for my youthful dark hair. Worrying about how my hair would look on this momentous occasion seemed hypocritical and shallow.

My daughter was planning her dream wedding overlooking a favorite beach. There would be family and friends—hers, his—the ocean, tent, candles, oysters, wine, sheep grazing in a nearby field, DIY wedding cake, my son's guitar playing, a gorgeous white handmade dress and golden belt, wildflowers . . . Part of my brain knew that it doesn't get any better than this.

The other part nagged, Wouldn't it be a little better without two-toned hair? Despite having recently read Dr. Christiane Northrup's book *Goddesses Never Age: The Secret Prescription for Radiance, Vitality, and Well-Being*, I didn't feel empowered. Instead, I fell back into the trap of imagining aging as decline and decay rather than as simply moving through time. It felt as if I were giving in to the whole going gray thing instead of giving myself over to a long but ultimately worthwhile process.

My lovely oldest child and I are close, sharing the renewable energy of unconditional mother-daughter love. We're highly attuned to each other, noticing vulnerabilities. But, unlike my mother and me, Lainey does not share the "see it, say it" gene. She is quieter, more reflective. While outwardly she can seem to be an easygoing, happy-go-lucky creature, she can also be diplomatic and can get cross when I overstep my boundaries.

Patiently, she listened as I bemoaned what to do about my hair. Turning the parenting tables around, Lainey comforted me, boosting my resolve. "What counts is on the inside, Mom."

"Choose a dazzling dress and your hair will look fabulous," Richard reassured me during a trim, while reiterating what everyone around me was also saying: "You don't want to leave the mother of the bride dress buying to chance."

The dress. I was so fixated on "the hair" that it was easy to forget "the dress"—the mother of the bride (MOB) dress. It would have to be eco-friendly, of course. My credo is the same whether it relates to food, fashion, or hair. What goes in and on my body needs to fit a clean and healthy ethos. Yet it is nearly impossible to grasp the magnitude of environmental damage the fashion industry creates in producing, packaging, and shipping clothing. Most companies don't have the resources to figure it out, either. Even if the fabrics are sustainably sourced, the clothing industry is laden with dyes, pesticides, and plastics, swelling our landfills, poisoning our air and waterways, exploiting workers, and wasting precious energy. As Annie Leonard says in her short film *The Story of Stuff*, there is no such thing as "away." When we throw something away, it must go somewhere.

Much of it goes into our waterways. When I learned that 83 percent of our drinking water contains microscopic plastic fibers from our clothing, I was not only grossed out but also worried about my family's health. Polyester, acrylic, nylon, spandex—they're all made from plastic. To add insult to injury, 89 million tons of oil was used in the textile industry in 2015. And almost all of the clothing we buy ends up in a landfill or gets incinerated, as a 2017 report shows in this depressing figure: eighty-one pounds of clothing per year, per person, is thrown out. It's enough to make you want to divest yourself of your yoga pants.

The MOB dress would, at the very least, have to check something off my squeaky green list. Minimal use of pesticides, chemicals, and energy? Supportive of fair wages and working conditions for the makers? I aimed for one out of five, knowing that it would not only be difficult but also, no doubt, cost a pretty penny.

According to a 2015 Nielsen study, a majority of consumers are willing to pay more for clothes that come from companies committed to positive social and environmental impact. I decided I would too—to a point. And it doesn't hurt that as a devotee of fashion magazines, I've always loved "the hunt" of shopping. But I live in upstate New York, which, with the exception of a few stores, is a fashion retail void. I shuffled what I thought was my not-so-bad-for-my-age body into the dressing rooms of my favorite local clothing boutiques to no avail. I was now a mere four weeks out from the wedding and ended up at an elite department store in Boston with my daughter.

I brought a few dresses to the mirror and layered them over my T-shirt and jeans. Nothing felt right. Without even looking at the fabric labels, I knew I would need to check my environmental ethics at the dressing room door. To make things worse, the mass-produced dresses seemed to have "Given up" written all over them.

I glanced at my daughter, who was shaking her head negatively at each outfit. I felt an emotional tsunami swelling up. When I indicated to Lainey it was time to leave, a saleswoman tapped me on the shoulder. Her badge read "Personal Shopper, Anne." When Anne offered her assistance, I pelted her with a tirade of MOB dress rules.

- No unnatural fabrics that could stand up and walk down the aisle on their own, that were created in sweatshops and then had to travel thousands of miles, burning away my grandchildren's future with polluting fossil fuels in transport.
- No ruffles or ruching in the middle, sides, or, God forbid, over the boobs—a dead giveaway something bad is hiding underneath.
- No dresses that could rival the resale value of my car.
- No pink. I'm not a pink girl.

- No sleeveless.
- And absolutely no colors that would clash with my hair.

I was expecting blowback from Anne, but instead she seemed unfazed. She disappeared into the aisles and returned loaded down with a candied kaleidoscope of shiny taffeta and polyester dresses. And to disguise any popover frump, many of these ensembles came wrapped in bunlike matching bolero jackets.

Just when I started to object, Anne pounced back and met my demands with her own harpy brand of MOB don'ts.

- Don't wear black—or the same color as the bride.
- Don't be too fashionable.
- Don't wear anything sexy.
- Don't wear a wrinkly fabric.
- Don't you know everyone will be looking at the bride, not you?

Getting in the last word, Anne seemed pleased with herself. With my shopping tug-of-war nerves shot, I caved and turned to leave.

Then Anne said, "I've noticed you're letting your hair go gray."

"I didn't 'let' my hair go gray. It grew gray. I'm no longer covering it up with hair dye," I snapped. I felt as if I'd crack under the heavy weight of old baggage.

"All righty, then, I prioritized your list and feel this dress is the one." Anne held up a blue-gray dress, streaked with gentle threads of silver, for my approval.

It fit like a glove. I didn't even ask if the silk was sustainably sourced.

Lainey took one look and said, "Mom, it doesn't get any better than this. It matches your hair. Take the dress. Now."

I was reminded of something singer Patti Smith once said. "I'm 67 years old. You're not going to tell me what to do. The only person who can boss me around now is my daughter."

It's easy to lose ourselves, relinquishing control to our adult children, whether it's a shopping decision or something more fundamental. Yet age comes with its own power—including purchasing power. Women sixty and older make up the fastest-growing group of consumers in the world, and we can vote with our wallets. While I can't fault any beleaguered brides or their mothers for jumping at the first dress that seems remotely plausible, what we buy does shape the way clothing is made. By the time Lainey is hunting for her own MOB dress, I hope the stores will be filled with beautiful, sustainable, and ethical choices—maybe even a dress that will match *her* elegant silver hair.

~

I can assure you that my husband, Ted, wasn't debating any ethical quandaries about wedding hair or outfits. Have you noticed that men age differently from women? For one thing, many men lose their hair. Score one for women. But believe me, you wouldn't find a man writing about gray hair, beauty, and age perceptions. Overall, men have an easier time dealing with aging in our society. In a national poll of 2,000 adults, nearly 90 percent thought women were under more pressure to look young than men. The poll also notes that men are considered old about five years after women. And here's the kicker: many people believe men get sexier as they age. Not so for women. Not only do women worry about becoming less attractive; they fear aging will hurt their careers. Forty-two percent of women aged fifty to fifty-nine felt that looking young helped them be successful at work, nearly double the number of men. Women also thought gray hair on men looked distinguished and that on women, it was associated with being old.

Ted assures me he doesn't share those views. Supportive of my gray hair journey, he jokes that he's developed a "graydar." With so many discussions about gray hair, he has started to notice bad dye jobs and "cosmetic enhancements," the off-kilter faces of celebrities who have had too much "work" done. Although Ted has been open-minded, even the most enlightened men approach women's beauty in surprising ways.

At a recent summer dinner party, I saw my husband deep in conversation with a male friend. I figured they were talking politics and was surprised when the friend asked, "How do you feel about Ronnie's gray hair?"

I was even more surprised when my earnest husband casually answered: "It's kind of like being with a blonde . . . a natural blonde."

Later that evening, I asked coolly, "What did you mean by the 'blonde' comment?"

"Oh, it's just guy talk," he replied.

Sure, women know all about "guy talk," I thought, remarking that no one has ever asked me how I like my husband's salt-and-pepper hair. When I hear "guy talk" between two liberal-leaning men who seem to have a firm grasp of the cultural #MeToo wave, should I give them a pass? These are not cavemen, nor are they cruel partners bragging about their sexual prowess or using name-calling to bump up their bully status. These are men somewhat confounded by sexual power imbalances, working their way through this latest cultural conversation. Notice I didn't say cultural "war." I have no interest in being at war with men, any more than I want to be at war with aging. Yet social conditioning creates a blind spot. Men get away with "guy talk" because they can. It seems to crackle from deep in their muscle memory. These men are not Trumps or Weinsteins, but they still yield power.

As much as I am saddened by our long history of sexism, I'm hopeful for change because women have power too. When we call guys on comments like these, many non-misogynistic men are willing and able to reexamine their words, as my husband did. When I pointed out my distaste for his sexist comment, Ted quickly realized it was not akin to a compliment; it was dismissive. He apologized.

For the sake of all the good men and the men we're raising, I believe this type of behavior comes from a fundamental lack of understanding. And, like any bad behavior, it can be learned from and corrected.

Speaking of the men we're raising, did my son, Ben, at twenty-five, notice the intense light-and-dark demarcation going on atop my head? As the only member of my family who had never seen my natural hair color (I started coloring before he was born), and as the only one who possessed the exact same natural hair color hair as I had when I was younger, he noticed—I think.

As I transitioned my hair, my son was experiencing his own series of transitions, from college to whatever comes next. During this period there were some intense emotional moments that led us to break through our self-obsessed shells, providing clarity on both of our transitions. Over time, I've discovered that Ben's random frankness about life is particularly refreshing. The most insightful comment I've received about my hair came from my seemingly indifferent son. Edging him out of his six-foot-six boy/man shell, fishing for something . . . a compliment, sympathy, I asked, "Does this gray hair make me look older?"

"Mom, true beauty is authentic. You got this. Keep it real."

"Real" is what the majority of men do. They let their hair do the natural thing, and many are solidly gray. Anderson Cooper, George Clooney, Steve Martin, Richard Gere, and Ted Danson all sported gray hair from the get-go as they advanced their careers and relationships on their distinguished looks. But not all men follow their lead.

According to the *Los Angeles Times*, the percentage of American men coloring their hair is on the rise, from 2 percent to 7 percent between 1999 and 2010, with 11 percent of men aged fifty to sixty-four now coloring their hair. Just For Men is the leading men's hair coloring brand in the United States. It saw recent sales of $137.3 million, a significant 50 percent increase in just five years.

In a *Men's Health* magazine article titled "Why Women Love Men with Gray Hair," men were encouraged to "manage" their gray hair with "*some* color." This men's fashion statement involves harsh bleaching, as ammonia combined with peroxide can leave virgin hair dry as a bone.

Taking a look at the ingredients of Just For Men hair color, you see little difference from women's hair dye. Yet there's a big distinction in the way men use color: many apply it to their beards and eyebrows. And the longer they use it, the more exposure they face from inhaling and ingesting irritants, and the greater their chance of developing asthma, allergic reactions, and other nasties, including cancer.

The Environmental Working Group's Skin Deep website rates the ingredients in Just For Men high and medium for such threats as endocrine disruption, allergies, and cancer.

Alarmingly, some younger men are taking a page from their counterparts' gray hair playbook as milky dyed gray hair is trending among younger women. According to a 2017 *New York Times* article titled "For Millennial Men, Gray Hair Is Welcome," an Amazon representative said the company has seen a threefold increase in searches for "gray hair dye" by men. The article suggested that hair dyeing is gaining traction with older men as well as millennial men. While some young men are dyeing to go gray, others are using "progressive" hair dyes—hair dyes with the active ingredient lead acetate, which is designed to gradually darken gray when used every few days.

When I learned that lead acetate is used in Just For Men, and also in Youthair and Grecian Formula (the latter, like Just For Men, also a Combe Inc. product), I was concerned. A group of consumer advocacy organizations, including the Environmental Defense Fund (EDF), Earthjustice, the Environmental Working Group (EWG), the Center for Environmental Health, Breast

Cancer Prevention Partners, Consumers Union, the Natural Resources Defense Council, and others, had already filed a petition with the FDA requesting that the agency ban the use of lead acetate, a known neurotoxin.

Lead acetate has been banned for almost a decade in Canada and the European Union. Health Canada found that "relatively small incremental exposures, such as those which would occur with regular use of hair dyes containing lead acetate, could result in the accumulation of potentially harmful body burdens of lead." The conclusion to ban lead acetate from progressive hair dyes was "based on data indicating skin absorption and possible links to carcinogenicity and reproductive toxicity."

Meanwhile, the FDA included this warning label on hair coloring containing lead acetate:

> Caution: Contains lead acetate. For external use only. Keep this product out of children's reach. Do not use on cut or abraded scalp. If skin irritation develops, discontinue use. Do not use to color mustaches, eyelashes, eyebrows, or hair on parts of the body other than the scalp. Do not get in eyes. Follow instructions carefully and wash hands thoroughly after use.

According to the EDF, these products were allowed to contain as much lead as 6,000 parts per million (ppm), even though the Consumer Product Safety Commission (CPSC) had banned the sale of household paint containing more than 600 ppm. Both the CDC and the FDA warn that no amount of lead in the blood is safe, as exposure can lead to serious health effects, especially in children. After using progressive hair dyes, men touch blow-dryers, combs, and faucets; the levels of lead found on these surfaces could harm children.

The countries that banned hair dye products with lead acetate have created new ones that are lead-free. The United States has some lead-free alternatives, but lead-containing products remained on the market until just recently.

While I was writing this book, the FDA announced it would repeal its 1980 approval for the use of lead acetate in progressive hair dyes. The EDF's chemicals policy director, Tom Neltner, cheered the move, saying, "[The] FDA's decision is an important step to protecting people from a continued source of exposure to lead that is a more significant route than the agency originally thought over three decades ago."

The FDA's final rule came down in October 2018. It amended lead acetate regulations and repealed approval for use of lead acetate in cosmetics to color hair on the scalp. "In the nearly 40 years since lead acetate was initially approved as a color additive, our understanding of the hazards of lead exposure has evolved significantly," FDA commissioner Dr. Scott Gottlieb explained in an agency news release.

"We now know that the approved use of lead acetate in adult hair dyes no longer meets our safety standard," he added.

The FDA agreed with the environmental groups' contention that since adult lead levels in the United States have decreased significantly since 1980, the agency's original conclusion that exposure to lead in hair dye was insignificant was no longer valid.

At the time of this writing, there is a thirty-day period for filing objections to the rule following publication. The FDA can begin enforcing the ban a year later—allowing industry time to reformulate products.

While lead acetate may be relegated to the dustbin of men's grooming history, PPD is still going strong. And that's a particular problem for African American men. A 2017 lawsuit alleges that the maker of Just For Men hair products, Combe Inc., "illegally targets African American men for its Jet Black hair dye." Citing a study by the Cleveland Clinic that showed black males are five times more likely to have a harmful reaction to PPD than whites, the plaintiffs' lawyers argued, "PPD is now known as one of [the], if not the most, common allergens in the African American population." The class action suit claims that even though the company knew the statistics, it used prominent athletes such as Emmitt Smith to market its products while failing to issue a warning about the danger to black men.

~

Only the young dye good? Like women, men are not only victims of our culture's obsession with youth; they've also been under the assumption that hair color is safe.

No matter our gender or race, we don't have to submit to the ideals created by the "beauty industrial complex." The use of dangerous chemicals in consumer products is a problem we can fix—both politically and personally—because we have power in the voting booth and at the beauty counter.

There's no doubt that jumping off the hair dye train will attract some sideways looks and comments. But it can be liberating to choose health. Who wants to waste precious time worrying about what others think or succumbing to inner doubts about beauty? For men and women, "good hair" can take you only so far in life. It's worth remembering, à la Brad Pitt's character in *Fight Club*, "You're not your fucking hair color."

Chapter 4

Romancing the Consumer

O N ONE OF MY MANY TRIPS back and forth to DC, I stopped at the Smithsonian Institution's National Museum of American History. The folkloric collections include early hair care products. In the late 1800s and early 1900s, hair dye was marketed to both men and women for "renewing" gray hair, along with cures for baldness and dandruff. I marveled at the dye ads from the nineteenth century, with their promises of covering up gray hair and beards with "natural" potions for "natural"-looking hair color. Dyeing one's hair was not common in those days and not well received in social circles. An 1877 *Scientific American* article mocked products such as Buckingham's Dye for the Whiskers and Ayer's Hair Vigor. At the time, doctors condemned hair tonics and dyes as quackery, warning that they contained toxic ingredients such as lead.

Archaeologists have traced hair dye back even further, to early civilizations. There's recorded evidence that around 1500 BC the Egyptians created a type of henna made from minerals, plants, and insects that was painted on hair with colors that carried cultural and social significance. Greeks and Romans created permanent black hair dye from plant extracts until they discovered it was too toxic. They even concocted a formula made from leeches. During the Roman Empire, prostitutes were commanded by law to have yellow hair, while "respectable" women would darken their hair with dyes made from a mixture of ashes from burned or extracted plants, nuts, and minerals.

By the time of the Renaissance, fair-haired angels were the rage. Using a mixture of alum, honey, and black sulfur, women spent time in the sun to encourage bleaching. Victorian women even wore large hats with open tops to allow their treated locks to soak up the rays. Gray was a popular color at the time, and hair powder was applied with silver nitrate, though overuse turned hair purple. As mentioned earlier, the era of synthetic dye began in the 1800s when chemists discovered para-phenylenediamine (PPD). They also found that hydrogen peroxide was a gentler, safer chemical for hair bleaching.

In the late 1800s and into the twentieth century, attitudes about hair color reflected Americans' anxiety about the number of immigrants pouring into the United States. Italian, Greek, Jewish, Russian, and other immigrants were maligned for being from an inferior "Mediterranean" race. They were considered dark and dangerous—not black and not white. Women with naturally light hair could separate themselves from the newcomers and claim a higher social status. Blonde was in vogue. Around this time, the stigma of looking old and gray took hold. Most American women kept hair coloring secretly under wraps. In the 1940s, hair salons actually had separate entrances for women who wanted to dye their hair, like celebrity hair salons have now. But all this was soon to change.

~

"It's a lot harder to be manipulated if you accept that there's a manipulator," marketing guru Seth Godin writes, "and it's a lot easier to see a path forward if you acknowledge that you weren't looking for one before."

Although Godin is known for his knowledge of marketing in the computer age, he could have just as easily been writing about the pre-feminist *Mad Men* era, when hair dye ads became omnipresent. As Godin says, a successful marketer is someone who understands the buyer, who "inspires us, delights us and brings us something we truly want."

After all, it was the dastardly and dazzling ad execs who shaped our mothers' hair coloring stories. And in the age of plastics and "better living through chemistry," those dashing shape-shifters transformed the word "natural" into a product placement misnomer. Advertisers understood that the key to selling hair color would be reassuring women that the dye was so natural that nobody would ever guess they were using it. This ushered in a modern era of hair dye.

In 1956, when the first home hair color kit was unveiled by Clairol, dyeing was redefined forever—making it "Nice 'n Easy" to color hair from a box. Copywriter Shirley Polykoff was the genius behind the campaign. Before Polykoff came along, the longtime stigma associated with dye was holding strong. Chorus girls and prostitutes changed their hair color, not nice girls. But Polykoff believed a woman should be able to be whoever she wanted to be, and she spun coloring as a secretive beauty choice. "Does she . . . or doesn't she? Hair color so natural, only her hairdresser knows for sure."

Finally, this undesirable sign of aging could be covered up with ease. Dyeing gray hair became a requirement, and sexism and ageism were not just tolerated, but expected. In fact, marketers aggressively pushed negative stereotypes about women, illustrated in Clairol's cringeworthy early print campaigns. Here is an excerpt from the 1943 original copy.

GRAY HAIR—THE HEARTLESS DICTATOR

Clairol swiftly, surely, secretly eliminates the heartaches of gray or graying hair!

Without justice or kindness, gray hair can rule your life. It can choose your clothes—confine you to a few subdued colors. It can pick your friends—from the "older set." It can dictate many things you say or do.

No wonder other women refuse to tolerate this tyrant. They prefer to have young-looking hair . . . "naturally with Clairol," the original Shampoo tint.

Smart women never accept substitutes—they insist on genuine Clairol. . . . Every trace of gray hair is covered permanently with tones so true and transparent they rival Nature's own.

What woman wants tyrannical hair?

Clairol bought ads in *Life* magazine in 1956, and Clairol sales skyrocketed almost overnight. Natural color shades were labeled with innocent names like Just Peachy, Tickled

Pink, and Frivolous Fawn, and new techniques such as "frosting" likened the bleaching of small strands of hair around the head to icing a cake. Marketers played on women's insecurities, offering hair products and procedures as surefire solutions to self-doubt.

By 1959, Clairol was considered the leading company in the US hair coloring industry. Over the next twenty years, "natural-looking hair color" became pervasive, with the number of American women who dyed their hair rising from 7 percent to 40 percent. Not only was Clairol's early campaign responsible for the dramatic increase in home hair color, but the company's TV and print ads also catapulted Clairol into iconic advertising history. Elated, women accepted this new beauty ideal, and with safety in numbers, there was nary a need to worry about the consequences of stigmatizing gray hair.

When Shirley Polykoff died in 1998, her *New York Times* obituary was notably titled "Ad Writer Whose Query Colored a Nation."

In the hit series *Mad Men*, adman Don Draper assures us with glaring conviction, "Advertising is based on one thing: happiness." Now, not only could women "privately" color their hair in the comfort of their homes, with "every trace of gray hair" covered, American women would be happy. Happiness sells.

～

Having grown up in the 1960s and early 1970s, happy is exactly how I remember my mother on "hair day." My suburban mom slipped away to have her hair done each week at the house of a neighbor who had set up a *Grease*-style beauty parlor in her avocado-colored kitchen. As a kid, I thought Mom's Friday mornings were reserved for top secret meetings straight out of the TV series *Get Smart*. I can't remember ever witnessing the tight rollers, hair dye, and hooded hair dryers that transformed her from housewife and mother to the glamorous groomed beauty she'd become. Notwithstanding, she would emerge Friday afternoon from her beauty portal ready for weekend action and smelling like Breck shampoo on steroids, with her teased, shiny brown, flipped-up shoulder-length bob looking identical to the one she'd sported the previous week. It was no mystery how her hair got styled into an unhuggable helmet that felt like clothes left out on the backyard clothesline to freeze: I observed Mom lacing her sweet-smelling hair with a noxious cloud of hair spray. As a child, I remember thinking Pepé Le Pew could sniff out my mother from inside the TV box and come out to flirt with a cartoon version of her. To this day, I have never seen a gray hair on her head.

But it wasn't my mom's midcentury hair adventures that first awakened me to the toxicity of hair color. It was my dad's brush with dye.

My dad was a "commercial artist." Today, his profession would be labeled "graphic designer," like my daughter's. Specifically, Dad was a wallpaper designer with an unrivaled sense of humor. His murals—flocks and modern metallics—were in full bloom all over the textured walls of my childhood home. My teenage friends would come over to "pet" the luxurious walls of my suburban Long Island cape. Dad's mod hand-printed silk-screening prowess was the envy of my extended family and their friends.

One weekend when I was about ten, my dad and I drove upstate to my uncle's house in Rockland County to paper the walls of his family's farmhouse bathroom. While Dad set up a makeshift wallpaper-hanging shop, I took advantage of being in the "country" and ran off to play by a stream with my girl cousins—a cheerful younger cousin and her moody teenage sister. When Dad finished papering for the day, he took off his paste-splattered glasses and went searching for a shower.

It took only one whiff of a newly wallpapered room to suspect that the toxic lowdown on wallpaper was mighty low. The chemicals used in vinyl wallpaper—polyvinyl chloride (PVC), lead, cadmium, chromium, tin, and phthalate plasticizers—were touted as miracle anti-mold protectants, while the adhesives in wallpaper paste emitted eye-tearing vapors that were later found to cause a plethora of health problems.

In fact, in 2010, researchers performed chemical tests on 2,300 types of wallpaper and found that many of the chemicals were linked to serious health problems, including asthma, birth defects, learning disabilities, reproductive problems, liver toxicity, and cancer. PVC plastic is two to thirty times more likely to contain hazardous chemicals than non-vinyl alternatives. Wallpaper is currently regulated under the Toxic Substances Control Act (TSCA). This Environmental Protection Agency (EPA) regulation has been severely flawed for several reasons. Most notably, instead of requiring that chemical companies demonstrate that their products are safe before they end up on store shelves, the law says the government has to prove harm first. The EPA can then regulate or replace the dangerous offender. Also, unlike every other major environmental law, the statute hadn't been significantly amended since it was adopted in 1976. In 2016, President Barack Obama signed the Frank R. Lautenberg Chemical Safety for the 21st Century Act. This amended TSCA to make it stronger and a more effective chemical safety system. Unfortunately, the improvements to TSCA have not been implemented by the current EPA, as the systematic weakening of the agency under the Donald Trump administration has been working to roll back and dismantle chemical safety rules.

So our nation's main bastion of chemical policy is unable to keep up with the changing science on toxic chemicals. I find this unimaginably sad. I've followed along for the past ten years as an incomprehensible political debate over regulating these chemicals has ensued, starring a strange string of Democratic and Republican bedfellows.

Therefore, it's probably a good thing that wallpaper hasn't made a strong comeback on the home design scene.

Dad was fastidious about getting the foul stuff off his body. When he emerged from the shower that weekend upstate, his brown hair was a bluish-black matte color. Thinking it was shampoo, Dad had washed his hair—and, he later confessed, his whole body—with my teenage cousin's intensely blue-black hair dye, which she used to match her raccoon-like black-lined eyes. I recall that he mentioned the funny smell of the "shampoo," which was possibly from the ammonia mixed with peroxide and dye. And, as I had never witnessed my mother getting her hair dyed, it was seeing that shock of black atop my dad's head that made me realize the power of hair color.

A few years later, when I was struggling with high school sciences and the dreaded New York State Regents exams, Dad chided me, "Remember, if it crawls, it's biology; if it doesn't work, it's physics; if it smells bad, it's chemistry." Thinking about Dad's mantra many years later, I wondered if a product that smells like it could cut through steel might not be the magical potion I wanted to use to make me look young and beautiful.

~

In my parents' era of "plastics" and "better living through chemistry," the adverse effects of toxic chemicals were hardly a blip on the environmental health horizon. But what came next was a shimmering sea change of activism. And women were at the helm of a movement.

In the late 1960s and early 1970s, with women's consciousness rising in so many arenas, beauty products started to target a new market: feminists. The slogan "Whatever makes you feel good is empowering" harked back to the happiness theory of advertising. And, of course, with feminist concerns firmly focused on a mosaic of political, economic, and social equality issues, there were much larger fish to fry than hair color.

Another woman copywriter, Ilon Specht, took on feminism and hair dye and ran with it. Specht worked with the French company L'Oréal to challenge Clairol's powerful hair coloring campaigns. L'Oréal's "Because I'm Worth It" campaign played on the idea that it's every woman's right to choose whoever she wants to be. And surrendering to ageism, baby boomers with big plans on the horizon feared that gray hair would surely be the quickest route to insignificance and invisibility.

L'Oréal's "Because I'm Worth It" campaign compared dyeing one's hair to exercising freedom of choice. Reinforcing that claim, the memorable 1968 Virginia Slims campaign jumped on the boomer bandwagon, targeting women with its slogan "You've Come a Long Way, Baby."

In Malcolm Gladwell's book *What the Dog Saw and Other Adventures*, he recounts the history of hair color marketing. When Gladwell spoke with Specht, she explained how L'Oréal came to upend Clairol's dominance in the hair coloring world in 1973.

> "I was a twenty-three-year-old girl—a woman," she said. "What would my state of mind have been? I could just see that they had this traditional view of women, and my feeling was that I'm not writing an ad about looking good for men, which is what it seems to me that they were doing. I just thought, *Fuck you*."

Specht sat down and wrote the copy in five minutes. She claimed the ad was very personal and that she was so angry that she didn't care about how much color cost. She cared about her hair, and not just the shade. She expected great color, but Specht also wanted her hair to feel "smooth and silky but with some body." That's why she wanted the most expensive hair color in the world: Preference by L'Oréal. Gladwell said Specht made a fist and struck her chest, exclaiming, "Because I'm worth it."

At the time, L'Oréal's Preference cost ten cents more than Clairol's Nice 'n Easy. In the 1980s, Preference surpassed Nice 'n Easy as the leading hair color in the United States.

Getting women hooked on smoking cigarettes aside, have we come a long way? In a recent *New York Times Magazine* article, "How 'Empowerment' Became Something to Buy," the author, Jia Tolentino, one of my favorite new writers, expresses her mistrust of using "female empowerment" to market products: "Women's empowerment borrows the virtuous window-dressing of the social worker's doctrine and kicks its substance to the side." Tolentino claims it's about pleasure, not power, and it's "individualistic and subjective, tailored to insecurity and desire."

~

With Clairol and L'Oréal marketing "natural" hair color in a box, the beauty torch was unceremoniously passed to a new generation of women. To a '60s generation who opposed what we believed was an unjust war in Vietnam started by old, gray guys, gray hair was not hip, nor was it valued. Our slogan was "Never trust anyone over thirty."

In 1970, we were also on the cusp of what would become the environmental movement, as the first Earth Day, on April 22, saw an estimated 20 million people nationwide protest environmental damage. Months later, President Richard Nixon worked with a bipartisan Congress to establish the Environmental Protection Agency (EPA), a new federal agency responsible for environmental legislation, ecological programs, and scientific research.

In the maelstrom of all these exciting fresh perspectives and passions, we lived with a growing infiltration of TV ads. Some admen understood the times were changing. David Ogilvy, later dubbed the "Original Mad Man," challenged the norm, writing, "The consumer is not a moron; she's your wife. Don't insult her intelligence. You wouldn't lie to your wife; don't lie to mine."

I clearly remember mocking Clairol's hair color message "The closer he gets, the better you look" as sexist. To my generation, the personal was political, and we wanted to be counted for more than our looks.

With cultural currents moving swiftly, baby boomers received two diametrically opposed and equally powerful messages about self-worth and beauty. On one hand, we internalized the idea that physical beauty should take a back seat to intelligence. On the other hand, the memo came through loud and clear that aging meant obsolescence. For the sake of our jobs and relationships, we needed to stay visible—vital, attractive, sexy—by striving for youthful beauty. Thus, we felt compelled to color our hair to hold off our own expiration date. Savvy marketers took aim and hit the mark, hooking us as they had our mothers, though the tagline of our hair stories was "natural," sans the bouffant.

~

In 1972, Gloria Steinem cofounded *Ms.* magazine. My friends and I read *Ms.* religiously throughout the '70s. With a passion for equality, *Ms.* led the pack, and Steinem's magnetism and enviable good looks crossed many boundaries, teaching us that being both beautiful and a feminist was okay. She also let it be known that traditional women's magazines shared a cozy relationship with advertisers that should not be trivialized. Advertising and editorial content were well coordinated, dictating how advertising dollars were spent. Clairol hair care and hair coloring products were one of *Ms.* magazine's first advertisers. This advantageous relationship for both *Ms.* and

Clairol allowed the hair company to be associated with the popular feminist movement, and *Ms.* needed national advertisers to add cachet, credibility, and cash to buoy up the new magazine. But the Clairol-*Ms.* marriage soon became contentious.

In a 1990 *Ms.* article titled "Sex, Lies, and Advertising," Steinem recounted Clairol's outrage at *Ms.* because, without warning, the magazine published an article that was pro–gray hair, citing the potential carcinogens in hair dye as a reason to stop coloring. This caused Clairol's ad agency to pull its advertising altogether for six months, and even after that, *Ms.* got "almost none of these ads for the rest of its natural life."

Steinem wrote:

> In the *Ms.* Gazette, we do a brief report on a congressional hearing into chemicals used in hair dyes that are absorbed through the skin and may be carcinogenic. Newspapers report this too, but Clairol, a Bristol-Myers subsidiary that makes dozens of products—a few of which have just begun to advertise in *Ms.*—is outraged. Not at newspapers or newsmagazines, just at us. . . .
>
> Meanwhile, Clairol changes its hair coloring formula, apparently in response to the hearings we reported.

Did advertisers care so little for women that they would treat them as guinea pigs, putting company profits before health? Steinem's opinion is clear:

> But the truth is that women's products—like women's magazines—have never been the subjects of much serious reporting anyway. . . . Though chemical additives,

pesticides, and animal fats are major health risks in the United States, and clothes, shoddy or not, absorb more consumer dollars than cars, this lack of information is serious. So is ignoring the contents of beauty products that are absorbed into our bodies through our skins, and that have profit margins so big they would make a loan shark blush.

Steinem ends "Sex, Lies, and Advertising" by asking, "Can't we do better than this?"

Not much better, it seems. As I came to discover many years later in that fateful meeting in Washington, DC, with environmental health experts, since World War II more than 85,000 new chemicals have become available in the United States that have never been fully tested for their toxicity to our health or our environment. And let's not forget that tongue twister para-phenylenediamine (PPD)—linked to cancer, allergies, and genetic damage in animals—which is a mainstay ingredient in most hair dyes.

∼

Have attitudes changed with the times? When Dove sent out a "Make Friends with Your Hair" survey, it found hair care was a higher priority in a woman's morning routine than eating breakfast, applying makeup, and getting extra sleep. And one in five women would pass on a social event if their hair didn't look right. In the cyclical nature of history, this seems eerily reminiscent of the "tyrannical" hair in those early ads.

British historian Mary Beard says women are victims of a great gray hair conspiracy by the hair coloring industry. Beard claims that in the past forty or fifty years the companies have amassed megaprofits by getting us to think that coloring is something women should just take for granted when they get older, "like a trip to the dentist." She also examines how the language has changed. People used to buy "hair dye." Now, in the United Kingdom they buy "rinses" and "tints," and in the United States we buy "highlights" and "colors." This all sounds joyous, fun, and even natural. Ah, the semantics of coloring. I've heard hairdressers say, "Dyeing is for Easter eggs—color is for hair."

The women of my generation, who famously rebelled against rules, wouldn't have anticipated we'd end up playing a dangerous game on artificial turf. Yet these same women—for years, myself included—live with a perennial quest to make their hair more youthful looking—spending tens of thousands of dollars over our lifetimes on salon visits, hair color, shampoo, and styling products.

With advertising agencies defining the look du jour and hair coloring companies getting little pressure from consumers or regulators to innovate, there is scant protection for women who consume hair dye as though it's a natural resource.

Even women who choose to go gray are encouraged to do it less than naturally. Consider the article "The New Over-40 Color Rules" in the recently defunct *More* magazine: "Your best gray may not be your natural gray." The article argues that "adding cool, ashy highlights can improve the color and is one of the best-kept secrets of stylish gray-haired women." *More* encouraged brown-gray semipermanent color, which covers some gray but not all.

This "best-kept secret," which is not a secret in "going gray" social media networks, perpetuates the old hair rule that women should cover up or improve their gray. The approach also ignores the reasons many women choose to stop coloring in the first place. There's certainly no mention of the continued "upkeep, the cost, the chemicals" in *More*'s story.

Even as many female baby boomers flourish in the arenas of work, relationships, and family, we continue to cling to hair dye, our health colliding with beauty ideals. Beard says the fear of gray is symptomatic of our fear of growing old and that everything's downhill after thirty. While I agree about the fear, I've also found that with age, it has become easier to let go of some oppressive beauty rules. This seems true for others too; a 2015 report in the *Journal of Personality and Social Psychology*, published by the American Psychological Association, found that self-esteem was lowest among young women but increased throughout adulthood, peaking at age sixty before it started to decline again.

~

As every ad exec knows—and as Don Draper implored as he swigged a Canadian Club—"If you don't like what's being said, change the conversation."

So let's change the conversation. Companies sell more than products. They sell values—concepts of worth, sex, success, and even health. They tell us who we should be. If companies won't live up to their moral obligation to see beyond their bottom line to keep consumers safe, consumers must demand to know what ingredients are in their stuff and whether or not they're safe. Gender and age equality can help us embrace confidence in gray hair and natural aging. We have a choice in how we define ourselves. That's *truly* truth in advertising.

And now one more from Seth Godin's arsenal of on-target marketing advice:

> You are not a brand.
>
> You're a person.
>
> A living, breathing, autonomous individual. . . .
>
> You have choices. You have the ability to change your mind. You can tell the truth, see others for who they are and choose to make a difference.

Chapter 5

Who's Looking Out for Us?

"WILL YOU ACCOMPANY ME TO THE SALON? It's time." Taking off my hat and scarf, I joined my dear friend Juliet for a lunch date at a local café. During my hair transition, we met weekly. Juliet happily commiserated with me each week as I shared my growing Pinterest "Silver Hair" inspiration board—it now displayed a curated collection of hairstyles in all stages and shades of going gray.

With her salt-and-pepper waist-length braid, and oodles of turquoise jewelry, she lived up to her cowgirl turned equine photographer persona. As she looked down at her hands, looped with silver rings, she repeated, "It's time."

A few weeks earlier, we had been at the same table in the same café when she leaned toward me, wiping tears that were falling on her glasses and into her tea, and cleared her throat. "My gynecological ailments have taken a turn. The ovarian cyst the doctors told me would be no more than nuisance surgery turned out to be cancer. I can't bear the thought of my hair falling out in bits and pieces. Would you come with me to the salon before the first chemo treatment?"

It all happened alarmingly fast. First, the braid got chopped, and then a razor buzzed off the rest. The next week, my colorful friend sat across the table with a super short pixie cut, an implanted access port popping out of her shirt, and talk of drugs, surgery, percentages, and recovery spilling from her pale, bluish lips. A few treatments in, she had no eyebrows or eyelashes, a slight limp from neuropathy in her foot, and the same sly smile.

I told her about an essay I had read in *Vogue* by a young cancer survivor, Suleika Jaouad, who said this about her ordeal: "Chemotherapy is a take-no-prisoners stylist."

"How does it feel to lose all your hair?" I asked Juliet, trying not to stare at her undecorated bald head.

Nonplussed, she shrugged. "I've been defined by my hair, and now I am defined by my cancer. The baldness is a symbol of saving myself."

My impulse was to ask how she would define herself once her hair grew back, but I held back my obsessive curiosity. Hair seemed trivial in the face of what she was going through.

A 2017 study by American Cancer Society researchers estimated that 40 percent of all cancer cases may be preventable, and nearly 20 percent of all cases are related to diet and physical inactivity. With close friends and family making agonizing health decisions, not surprisingly, I've noticed women wringing their hands and spirits about whether they could have done anything to prevent their diseases.

That questioning goes along with a new sense of hyper-vigilance that throws people into a cycle of relentless uncertainty about their health. Looking back to move forward, they try to make sense of the senseless. It goes something like this: "Maybe it's my diet? The bacon? The sugar? The wine? Oh, why didn't I bite the bullet and buy organic fruit and vegetables?" And this: "Maybe if I had used cleaner beauty products, without questionable chemicals and dyes?" *there is, Deuteronomy 30:15-20*

Of course, there is no way to know whether taking these kind of precautions could have saved an individual from a heartrending cancer diagnosis. But the anecdotal evidence is hard to discount. When Jackie Kennedy died of non-Hodgkin's lymphoma, many pointed to several studies that found dark hair dye increased the risk of the disease. In fact, for months after her death, in May 1994, Manhattan salons saw a dip in clients coloring their hair in dark shades. It was also whispered that Elizabeth Taylor's brain tumor was caused by her continuous use of dark hair dye.

Rumors, maybe. But the US Department of Health and Human Services' National Toxicology Program (NTP) classifies some chemicals in hair dye as "reasonably anticipated to be human carcinogens." When government agencies use this language or say that a chemical is "likely carcinogenic," they are making a distinction with chemicals that are "known carcinogens." Epidemiological studies, which look at cancer rates in human populations, and lab testing of animals are both used to determine how particular substances are classified.

In this case, dark dyes contain more coloring agents than lighter shades, including "some chemicals that may cause cancer." The National Center for Biotechnology Information (NCBI), part of the National Institutes of Health (NIH), cites these dyes as a major risk factor for bladder cancer. The NCBI found that women who used permanent hair dyes at least once a month had twice the risk of bladder cancer than did non-dyers. The NIH also concluded that salon workers who worked for ten or more years as hairdressers or barbers experienced a fivefold increase in cancer risk compared with those not exposed.

With such alarming statistics and more than 75 percent of women using hair dye, one has to ask about pregnancy, as the womb is its own incubator. Not surprisingly, studies have found that exposure to toxic chemicals during pregnancy and lactation is omnipresent. The National Cancer Institute (NCI), also part of the NIH, notes the chilling reality that "babies are born 'pre-polluted.'"

Studies published in the NCBI's *Journal of Clinical and Aesthetic Dermatology* point to a link between use of hair dye during pregnancy and the development of several childhood malignancies. Babies are particularly vulnerable, not only because they're still developing but also because they're physically smaller than adults, which means the same amount of toxin packs more punch. The NCBI concluded that concerned pregnant women should avoid all hair coloring.

Hair dye is complex; it may contain any of thousands of different chemicals, and each product includes a different cocktail of ingredients. The products have also changed over the years, so the cumulative effects are not standard. But the risk factor remains. So does the unnerving question *Who's looking out for us—and for our legacy, our children?*

If you were like me before I got sucked down the rabbit hole of chemical reform, when you made a decision to put on lipstick or nail polish or to color your hair, you gave it little thought. You were simply anticipating that feel-good pleasure of a beauty transformation, assuming that something or someone—a government agency, you hoped—had made an independent evaluation of your product and determined that it was safe.

It's on the shelf. This is America. It must be safe. Unfortunately, that's not the case. This is why there's no easy answer to the question, Should I dye my hair or not?

For the most part, the government doesn't determine whether or not our products are safe. What it does is set a framework for businesses to follow and for consumers to follow suit.

Although the Food and Drug Administration (FDA) didn't receive its current name until 1930, the agency began its regulatory functions with the passage of the Pure Food and Drug Act of 1906. With this new law, it became illegal to misbrand or adulterate foods, drugs, and drinks in interstate commerce. But producers and manufacturers did not have to submit information to the FDA (then called the Bureau of Chemistry) before marketing their products. The burden of proving a food or product was mislabeled fell on the government. This turned out to be immensely difficult. When 107 people were poisoned and died from Elixir Sulfanilamide, a "medicine" that basically consisted of antifreeze, the public urged Congress to pass legislation that led to the Federal Food, Drug, and Cosmetic Act. President Franklin D. Roosevelt signed the FD&C Act into law in 1938. This law required manufacturers to demonstrate the safety of their products before they could go to market.

The question is, What does it mean to demonstrate product safety, and who makes sure the companies are really doing it? While the FDA can regulate cosmetics, it does not preapprove products before they land on store shelves. That is mostly left to the Cosmetic Ingredient Review (CIR): a program of the Personal Care Products Council (PCPC), a trade organization that represents the US cosmetics industry. The CIR's purpose is to assess the safety of ingredients used in cosmetics. It consists of a panel of scientists and three nonvoting members from the FDA, the Consumer Federation of America, and the PCPC.

Jay Ansell, vice president of cosmetic programs for the PCPC, says that risk management really comes down to three entities: the FDA, the CIR, and the companies. He acknowledges, "The primary responsibility of assessing safety falls on industry. Companies, before they put something on the market, are obligated to assure the product is safe. That assertion is validated by the CIR assessment."

Here's the kicker: the PCPC established and funds the CIR. In other words, it is the responsibility of the product manufacturers to decide whether or not the ingredients in their products are safe according to FDA rules. The conflict of interest is stunning. By leaving safety oversight to the CIR, we are essentially letting the fox guard the henhouse.

The only check on the CIR is that the FDA can take action after a product is on the market if it is found to be either in violation of the law, which generally means it's been mislabeled, or harmful. The agency can then get a federal court order to stop sales, request that US marshals seize the product, or bring criminal charges.

While the US government is not testing individual products before they go to market, it does conduct toxicology research. The National Toxicology Program (NTP) was established in 1978 to coordinate toxicology research and testing across the Department of Health and Human Services, including the NIH, the Centers for Disease Control and Prevention (CDC), and the FDA.

In this tangle of agencies, councils, and panels, one needs a glossary to keep up with all the acronyms. Yet, even with all the government bodies involved, the FDA, with its annual budget of just $8 million, has twenty-seven staff members and only six inspectors for approximately 3 million foreign shipments of cosmetics that come into the United States annually. This includes lipstick, hair dye, tattoo ink, nail polish, and other beauty products.

"The FDA believes that the vast majority of cosmetic products on the market are safe, but our information on the universe of cosmetics is limited," Linda Katz, director of the FDA's Office of Cosmetics and Colors, told the *New York Times* in a written statement.

So what does the alphabet soup of regulatory agencies say about hair dye? The NTP has not classified exposure to hair dyes as a potential cause of cancer. As mentioned, it has, however, classified some chemicals in hair dyes as "reasonably anticipated to be human carcinogens." One chemical of concern is para-phenylenediamine (PPD), which is widely used in hair dyes. Animal studies have indicated that it may be carcinogenic, and it is the main culprit when it comes to hair dye allergies.

Yet this classification does not mean that the FDA will restrict these ingredients—or even that it has the right to. As the American Cancer Society (ACS) notes, when the FDA was created, some ingredients that are still used today in hair dye were excluded from regulation.

One big fat exclusion was coal-tar hair dyes. While the FDA regulates most color additives, "coal-tar colors" got a pass as by-products of the coal industry, though today the dyes are made from petroleum. Because of some confusing regulatory structure created by the FDA and the FD&C Act, the "coal-tar" name is still used in permanent, semi-permanent, and temporary hair dyes. This further muddles the question of what is safe and what is harmful.

According to the agency itself, "FDA's ability to take action against coal-tar hair dyes associated with safety concerns is limited by law." In the 1980s, the FDA did publish a rule requiring a special warning statement for all hair dye products containing these ingredients: 4-methoxy-m-phenylenediamine 2,4-diaminoanisole and 2,4-methoxy-m-phenylenediamine sulfate 2,4-diaminoanisole sulfate, because "some coal-tar hair dyes were found to cause cancer in animals."

The cosmetic industry reformulated coal-tar hair dye products, and the FDA continues to collect "adverse event data" to determine if this class of ingredients is safe. In other words, the agency asks consumers to report bad reactions to hair dye (and to any other cosmetics)—after getting any necessary medical help. The FDA has to rely heavily on these reports because hair dye manufacturers are not required to share their safety data or consumer complaints.

Even though companies tweaked their ingredients to avoid the special cancer warning, they are still using related chemicals that can cause allergic reactions, including irritation and hair loss. This is where para-phenylenediamine (PPD), the most common component of hair dye, comes in. The FDA warns that coal-tar hair dyes "are not intended to be used for staining the skin. . . . It's important to follow the directions on the label." Yet many dark tattoo formulas contain PPD.

Instead of restricting PPD or all coal-tar dyes, the FDA requires a product caution I'd wager most salon clients have never read: "This product contains ingredients which may cause skin irritation on certain individuals and a preliminary test according to accompanying directions should first be made. This product must not be used for dyeing the eyelashes or eyebrows; to do so may cause blindness. (FD&C Act, 601(a))."

The FDA also issued this nifty Coal-Tar Hair Dye Safety Checklist:

- Follow all directions on the label and in the package.
- Do a patch test on your skin every time before dyeing your hair.
- Keep hair dyes away from your eyes, and do not dye your eyebrows or eyelashes. This can hurt your eyes and may even cause blindness.
- Wear gloves when applying hair dye.
- Do not leave the product on longer than the directions say you should. Keep track of time using a clock or a timer.
- Rinse your scalp well with water after using hair dye.

- Keep hair dyes out of the reach of children.
- Do not scratch or brush your scalp three days before using hair dyes.
- Do not dye or relax your hair if your scalp is irritated, sunburned, or damaged.
- Wait at least 14 days after bleaching, relaxing, or perming your hair before using dye.
- Read the ingredient statement to make certain that ingredients that may have caused a problem for you in the past, such as p-phenylenediamine (PPD) are not present.
- If you have a problem, tell your healthcare provider. Then, please report it to FDA.

First of all, who knew there could be such a thing as a Coal-Tar Hair Dye Safety Checklist? I had never seen this list, and I'd guess neither had my various hair stylists over my two decades of coloring.

Here's what raced through my thoughts: Never had a patch test. Hair dye dripped into my eyes and caused stinging many times. Had no idea how long the dye needed to stay on my hair. My colorist said my gray was getting resistant and the dye needed to be left on longer than recommended. For days after coloring, when I scratched my head, my scalp would itch and dye residue would stain my fingernails. But that wasn't the worst of it. Over the years, I had developed migraines that seemed to be triggered by hair dyeing sessions.

Reluctantly, I mentioned having a possible adverse reaction to hair dye to my doctor twice. The first was when I was in a seemingly endless migraine headache cycle. The second was when I had a facial rash that started near the corner of my eye and was spreading. The doctor said the migraines were probably caused by hormonal shifts. She prescribed migraine medication and asked if I would consider taking bioidentical hormones for perimenopausal symptoms. I picked up the prescription for Imitrex and ordered the hormones. Imitrex worked on the headaches but left me spacey, spinny, shaky. I never took the hormones.

The rash raged across my face like wildfire, so my doctor gave me a steroid cream and told me to stop wearing makeup. Unhappily, I complied. The cream made the rash worse; my skin erupted in red, itchy bumps. So she sent me off to a dermatologist, who changed the steroidal concoction. The rash continued its fury, oozing dangerously close to my eyes, nose, and lips. Despite it all, I desperately wanted to wear makeup to cover up the mess. Instead, I canceled social engagements right and left.

On the urging of a friend, I went to see another local doctor. He was a doctor of internal medicine who was also a classically trained homeopath. After an hour-long discussion of my medical history, including my diet and lifestyle activities (and inactivities), two cesareans, multiple Lyme disease infections, and the dreaded migraines, he told me the rash was probably caused by an irritation from a chemical compound in one or many of my personal care products. He suspected it was SLS—sodium lauryl sulfate. SLS is the first or second ingredient in most toothpastes, shampoos, and foaming cleansers. I rooted around and found that SLS was having a field day in my bathroom. It was the second ingredient in my "healthy" Tom's of Maine toothpaste. The expensive Bumble and Bumble shampoo that made my dyed hair shine like a polished onyx gem-stone had it. And it was included in the convenient "relaxing spa aromatherapy" pump bodywash.

Sodium lauryl sulfate (SLS) and sodium laureth sulfate (SLES) are strong surfactants that can cause skin irritation or trigger allergies. In fact, the homeopath told me SLS is used in lab tests to purposely irritate the skin's outer layer. I learned that SLES is often contaminated with 1,4-dioxane, a by-product of a petrochemical process called ethoxylation, which is used to make other chemicals less harsh.

Like PPD and formaldehyde, SLS is banned in the European Union. Health Canada has categorized SLES as a "moderate human health priority" risk. It's currently up for future assessment by the Canadian government's Chemicals Management Plan.

"Don't ignore your body's subtle signals. Your body is telling you something." The kind homeopath said that my face was suffering from "bubble trouble." I had to ditch products containing SLS immediately. He suggested I try using these three guidelines when choosing personal products:

1. Look for products that have no more than three to five ingredients.
2. You should be able to pronounce the ingredient names.
3. Don't wash your face more than once a day, and don't wash your hair every day.

Coincidently, at that time I was reading Michael Pollan's slim but impactful book *Food Rules*. Pollan's guideline for food seemed to mirror a parallel universe. "Avoid food products that contain more than five ingredients. The specific number you adopt is arbitrary, but the more ingredients in a packaged food, the more highly processed it probably is."

As with food, it is not so easy to find beauty products with just a few ingredients and avoid those we can't pronounce. Being aware of body pollution is half the battle. The other half is changing behavior.

Anyway, I can't be sure whether or not SLS was the cause of my "bubble trouble." But I cut SLS out of my beauty routine, and the rash disappeared and never returned.

In all of this crazy-making, I recommitted that I would never go back to coloring. After learning that of the 4,500 ingredients reviewed by the CIR in the past forty years, only 13 have been deemed unsafe, I shut down the hair dye party for good. After all, that particular party could only end on a sour note—one that could affect millions of women and men. It seemed unconscionable that while Europe had banned or restricted 1,400 chemical ingredients, the US government failed to act on all but 13.

Since the FDA can never be absolutely certain of products' safety because of the "limitations of science," the door is left wide open for the chemical lobby. The Trump administration has dusted off a "Make US dirty again" welcome mat and is living up to threats to cut federal funding for science and the environment. In the name of protecting companies from "evil" regulations, the administration has failed to protect its citizens and future generations.

A while back, I wrote an article titled "Don't Let Politicians Pollute the EPA" in response to a proposal to eviscerate the Environmental Protection Agency. The image I chose for the article came from Documerica, a project by the EPA that collected photographs showing the effect of environmental issues on everyday life in the 1970s. One photo from 1973 (the year I graduated high school) captured dark, hazy smog hanging dangerously low over the George Washington Bridge in New York. Across the country, images like that one had come to define our environment. Horrific air quality and polluted waterways, some of which were actually catching on fire, ignited the public consciousness. In a collective bipartisan roar that transcended politics, legislators and their constituents alike vowed to not let Earth be destroyed by pollution.

Our environmental protections have come a long way since the 1970s, helping to drive down pollution and improve the health of Americans. But while we've made tremendous scientific strides in addressing air, water, and body pollution, we are currently in danger of reversing course.

Today, the federal government is challenging science itself, with agencies eliminating or changing scientific language on their websites, and purging grants, personnel, and scientific advisory boards. Most disturbingly, the administration is delaying and decimating health regulations that protect the most vulnerable. President Trump even signed an executive order stating that for every new regulation, two existing regulations must be repealed. The maniacal justification for this "screw you" deregulation bender is that it will save billions of dollars. The question is, save *who* billions? Certainly not the American workforce, who will spend more on health care and become less productive on the job. This two-for-one crackdown spelled a bad deal for those who can least afford the health and economic burden. And Trump has made good on his promise.

The Trump administration's assault on public health beats us down, threatening to leave a long-lasting scar on our nation. But even before Trump came into office, the United States lagged far behind Europe in the area of chemical reform; Europe is now moving toward testing chemicals before they go on the market. Meanwhile, we have become sitting ducks for the chemical lobby. Yet the majority of Americans support government agencies that put citizens' health over polluters' bottom lines. At Moms Clean Air Force, we like to say, "No one voted to make America dirty again."

Nicolas Kristof, in his *New York Times* column, equated what's happening today in toxic chemical research to what happened in the 1950s with cigarettes. At the time, researchers were discovering that cigarettes caused cancer. Someone in the United States was dying every minute and a half as a result of smoking, according to health officials. The tobacco companies, rather than examine their products, spent a considerable amount of time and money to discredit the research. By the early 1970s, President Richard Nixon had caved to pressure from the tobacco lobby and considered eliminating funds from the federal budget for future health assessments. This was prevented only by a last-ditch effort on the part of advocates who fought the cuts with science-based campaigns that proved industry should not have the right to promote harmful tobacco without public oversight.

Looking through this political viewfinder, it is easy to see commonalities between the Nixon era and today. In 2015, the chemical lobby spent the equivalent of $121,000 per member of Congress. Now we have a president who scales back protections when heavily lobbied by the chemical industry, claiming there is little risk from chemical exposure. In the 1970s, advocates fought back and won—the question is, Will we?

Author Terry Tempest Williams writes that we are once again at a crossroads: "We can continue on the path we have been on, in this nation that privileges profit over people and land; or we can unite as citizens with a common cause—the health and wealth of the Earth that sustains us." If we cannot hold polluters accountable, "then democracy becomes another myth perpetuated by those in power who care only about themselves."

I for one don't believe democracy is irrevocably broken. It's a scary time, for sure. It seems like the government is attacking itself and the casualties will be the American people. Yet not every politician marches to Trump's drum. Right now, legislation is being proposed that could help protect consumers from dangerous chemicals—if Congress summons the political will to be responsible to its constituents.

The way to change the game is to change the law. Senators Dianne Feinstein (D-CA) and Susan Collins (R-ME) and Representative Frank Pallone (D-NJ) are doing just that as they work together to create the Personal Care Products Safety Act. This bipartisan bill is designed to create some semblance of consumer protection, mandating that the FDA have more oversight over the safety of beauty products.

Senator Feinstein told *New York Magazine*'s The Cut, "From shampoo to lotion, everyone—women, men, children—uses personal-care products every day. Despite the universal use of these products, none of their ingredients have been independently evaluated for safety. This puts consumers' health at risk and we urgently need to update the nearly 80-year-old safety rules."

Here's a summary of the mandates of the Personal Care Products Safety Act:

- The FDA would have to evaluate a minimum of five ingredients per year for safety.
- The FDA could order recalls of questionably safe beauty products.
- It would require complete ingredient information for products to be available online and printed on labels.
- The FDA could require products that include ingredients not suitable for children to be labeled.
- The FDA could issue regulations on Good Manufacturing Practices as they relate to beauty products.

What could this mean for hair products and potential health offenders such as PPD? It could change coal tar's protected status. As mentioned, coal tar is a carcinogen, but current law prevents the FDA from restricting it so long as products include a warning label.

A number of large cosmetics companies, including L'Oréal, as well as health and consumer advocacy groups, support the bill. Some companies are opposed because they predict the stringency of the proposed legislation would unfairly penalize smaller independent cosmetic companies.

This bill is a big step in the right direction, and I applaud the bipartisan efforts of Senators Feinstein and Collins and Representative Pallone as well as the companies willing to support safe policies over their bottom line, making our health a priority. The Personal Care Products Safety Act would give the FDA the power to review controversial chemicals and determine whether those ingredients are safe or unsafe.

And, as I've learned through years of advocacy work with Moms Clean Air Force, there is much we can do to help our elected officials do the right thing. Arming consumers with sound science, like applying the lessons of the tobacco advocates, gives citizens the power to demand that our country's legislators not sit idle while the beauty industry is allowed to put coal tar, hormone-disrupting phthalates, cancer-linked parabens, and just about any chemical in our everyday products. We can also boycott products to pressure companies to disclose their ingredients. As my father used to say, "We can raise a stink."

So, for now, we're left self-regulating. This has us sleuthing around, reading labels, asking more questions, and figuring out answers. It leaves us floundering in the land of "Let the buyer beware." There's virtue in being a savvy consumer, but in the era of archaic laws, industry panels, deregulation, and a powerful chemical lobby, self-regulation goes only so far. We need reasonable regulation and also more scientific research to learn about health risks. Unfortunately, a financial bump to the FDA's budget doesn't seem forthcoming. So we're left second-guessing our choices, wondering if that cocktail of chemicals we applied for decades may have left an invisible mark—the kind of stain cancer could prey on.

~

Juliet is now more than five years away from cancer's brink. I ask her if she is still defined by her diagnosis of ovarian cancer. It was grueling, she says, physically and emotionally painful; she's had to mastermind her way through the labyrinth of the health-care system. And yet, with a conscious decision to keep her new crop of hair defiantly short, she seems almost serenely calm, instinctually driven in her purpose to build a new business from a passion she's had her whole life. Juliet's face softens when she answers, "I'm defined by being a cancer survivor. I feel wiser and happy because people who don't die get to grow older."

Chapter 6

Dumping Dye Down the Drain

I AM FIDGETY BY NATURE. I've been aware of this since third grade, when my teacher, Mrs. Betz, ordered me to sit stiller, quieter, and longer than the rest of the class. Eventually, I figured out how to stop squirming and be quiet in situations that required intense focus. I became a master at it in my late teens when I learned to knit.

Knitting has made life immensely more livable. In waiting rooms; on planes, trains, ferries, and long car rides; during conference calls; and while watching Disney movies years ago with the kids (I carried a teeny-tiny flashlight for the theater), my needles clicked away productively.

I even spun knitting into a job in the 1980s after winning a *Family Circle* magazine knitting contest. I designed patterns and sold handmade sweaters. Now there's a lucrative profession!

While my career in yarn was short-lived, to this day knitting quiets my brain. It allows me to click into a contemplative, observational space that I can't always access when I am antsy. I'm at my best when I am knitting because I've learned how to sit with myself. That patience served me well during the long days, weeks, and months I spent waiting for my dyed hair to grow out.

Knitting also gave me an early glimpse into the dangers of dyes. Challenging myself to up my game, I took a hand spinning and yarn dyeing class. The class began with a chemistry lesson, as we were using artificial as well as natural dyes.

"The chemicals used in the yarn dyeing process can include toxics, acids, bases, and oxidizers," the fiber arts teacher explained as we put on gloves to mix the chemicals used to dye our hand-spun yarn. "To prevent chemicals from seeping into our air, water, and soil, the liquid dye bath needs to be disposed of properly."

My lab partner asked, "Can't we just dump the leftover dye down the toilet?"

"Most yarn dyes are considered hazardous waste," the teacher quickly answered, as if she'd been asked this a million times.

Whether we're using it to color our sweaters or our hair, dye poses a real hazard to the environment. Along with formaldehyde, nail polish, paint stripper, oven cleaner, and a multitude of other household and personal care products, the dyes that get dumped down the drain don't disappear when they swirl out of sight.

More than an expertise in hand dyeing, the lesson I took from my knitting class was that chemicals in everyday products eventually come full circle, ending up in our air and water. It was reinforced when I read *Cradle to Cradle*, Michael Braungart and William McDonough's brilliant book about the life cycles of products and environmentally friendly design. Strands of knowledge like these wove themselves together in my brain, creating a new understanding of how we can make things ethically, with reverence, an understanding that became a passion and ultimately a career.

Back when I was a teacher, I would knit my way to calm at long faculty meetings. I remember, during an especially contentious exchange among a community of beloved fellow teachers, clicking away at a complicated stitch pattern. As the bickering hit a maxed-out tenor across the conference table, my colleague and mentor, Eric Tomlins, glanced over at me as I silently ripped out two full hours of knitting. I needed to start over, casting on, knitting, purling, cabling, with yarn I had spun and dyed myself.

Finally wrapping up the faculty meeting, Eric said something to this effect: "If we want to get to the heart of a problem, we're going to have to first unravel it and follow the situation back to the place where we made the mistake. Just take a look at Ronnie . . ." All at once, a long table of tired teachers turned their chairs to examine my hands. Eric continued, "We all must become master unravelers, connecting and reconnecting the dots, and continue to work while things take shape."

I learned much from my twenty-five-year friendship with Eric, but the most important thing he imparted, when he was dying from amyotrophic lateral sclerosis, ALS, was to follow my passion. While I may lack in attention, I'm told, I have passion in spades.

Eric's untimely death at fifty-four stopped me dead in my tracks. It left a gaping hole in my heart. For a while, it seemed like that dropped stitch would be too hard to repair. But he knew me well, and right before Eric passed, I confessed to him that I couldn't keep on teaching without him at the helm. It was his enthusiasm for my work that kept me afloat in the job. With great effort, in his new breathy speech pattern that just a few months earlier had been succinct and clear, Eric said, "You have another work life in you. Follow that passion."

And just like that, he guided me out of my comfortable, but stifling, teaching nest and into my next adventure.

After a series of fortunate events, I found my new tribe, the tireless Moms Clean Air Force team of women (and one man). I love helping to activate parents to protect their families from air pollution and toxic chemicals. Just as Eric said, finding solutions often means working backward, following threads, until we discover the source of the problem. In the case of environmental health, the problem is too often of our own making. If we want to protect ourselves and our world from toxic chemicals, we have to examine and rethink the products that contain them. And that may require some ripping out and reknitting.

~

Early in our work on air pollution at Moms Clean Air Force, we came across a digital map designed by the folks who lead Google's Big Picture visualization research group. The map allows the viewer to watch, in real time, the pattern of wind moving across the United States. Dominique Browning, Moms Clean Air Force's cofounder and senior director, wrote: "Wind connects us all. It is a huge force, moving in such enormous sweeps across our skies, that it overwhelms the boundaries we have thrown up, the ways in which we measure where we are—backyards, city limits, state lines."

I was reminded of how we share the air when I recently came across this headline: "New Study: Daily Emissions from Personal Care Products Comparable to Car Emissions, Contribute to Air Pollution in Boulder."

Could chemical emissions from our products, our very own "personal plume," contribute to climate change?

We've all walked past an overly perfumed woman and reflexively brought our hand up to shield our nose from the artificial and often acrid smell. That's one obvious example of how the personal plume of pollution affects others. But it turns out those chemicals do more than simply stink. According to a 2018 study by the Cooperative Institute for Research in Environmental Sciences (CIRES) and the National Oceanic and Atmospheric Administration (NOAA), a common ingredient found in shampoos, lotions, and deodorants emits pollution that is similar to the exhaust coming from cars and trucks.

How'd they figure this out? The scientists took measurements during rush hour in Boulder, Colorado, to track the concentrations of traffic-related chemicals, including benzene, a marker of vehicle exhaust. In the soup of about 150 chemicals swirling in the air, one stood out because the scientists found a big peak in the data but didn't know what caused it.

Siloxane (decamethylcyclopentasiloxane) peaked in the morning, along with benzene. The group thought it might also be a chemical in vehicle exhaust. But when the scientists tested tailpipe emissions directly (phew), they couldn't find siloxane. It turns out the chemical was coming not from commuting cars but from the squeaky-clean people inside them. Siloxane is actually a silicone-based ingredient used in cosmetics, deodorants, defoamers, lubricants, and soaps to soften, smooth, and moisten.

"We detected a pattern of emissions that coincides with human activity: people apply these products in the morning, leave their homes, and drive to work or school. So emissions spike during commuting hours," says Matthew Coggon, a CIRES scientist.

In addition to making our hair and skin soft and silky, siloxane belongs to a class of chemicals called volatile organic compounds (VOCs). Although siloxane evaporates quickly once it's applied, when VOCs hit the air, sunlight can trigger them to react with nitrogen oxides and other compounds to form ozone and particulate matter. Ozone and particulate matter are types of air pollution regulated by the EPA because of their effects on air quality and health.

One type of siloxane, cyclotetrasiloxane (also known as D4), is a particular favorite of hairdressers; it's used in conditioning products to coat the hair strand. Along with its cousin, cyclopentasiloxane (D5), D4 is also found in soaps, baby bottle nipples, cookware, and home cleaning products. The Environmental Working Group's database of ingredients in personal care products found D4 and D5 in more than 1 of every 7 products. Notably, hair care products with D4 and D5 toxins were found to be extremely persistent in the environment, building up in the food chain and potentially in people's bodies. The toxins were found to travel thousands of miles by air.

Canada is so concerned about D4 and D5 siloxanes that it will prioritize them for action under the government's Chemicals Management Plan. Not so in the United States. As the CIRES and NOAA study points out, while transportation emissions of VOCs have decreased because of stricter air pollution regulations, other sources of VOCs, including those from personal care products, have emerged as contributors to urban air pollution. The research team is now looking at other chemicals in our personal plume that affect air pollution.

~

What goes up must come down. Pollution floats through the atmosphere and dives deep into the water. As we pump carbon dioxide into the air, it is absorbed by the oceans, making the water more acidic, which makes it harder for creatures such as oysters, corals, and mussels to form shells. And as climate change leads to more extreme weather and heavier rainfalls, more runoff will pollute our waterways. This is all aside from the chemicals, including those in hair dyes, that are flushed down the drain and eventually end up in our rivers, lakes, and oceans.

With the oceans covering more than 70 percent of the planet and containing 97 percent of Earth's water, H_2O is the basis for all life. And whether we live next to the ocean or miles from any waterway, clean water is critical for survival.

Marine biologist Sylvia Earle knows the importance of water. She has spent more than fifty years scuba diving and doing scientific research in pursuit of understanding the oceans. Sylvia's conclusion is ominous: "The oceans are dying."

When asked in a 2014 *National Geographic* interview after the BP oil spill whether or not she thinks people have become complacent about ocean pollution, Sylvia responded:

> We could not have seven billion people if we were still burning whale oil and wood. Our prosperity is deeply rooted in using energy from coal, oil, and gas. It has given us so much wealth and benefits, but now we have to shift gears and go to something else, because the downside of burning fossil fuels is threatening our life-support system. . . .
>
> We must not lose sight of the real cost of the terrible legacy we may leave ourselves.

One of the reasons Sylvia believes the oceans are dying is that plants and animals rely on oxygen to survive. Warmer water holds less oxygen. So does water polluted by industrial plant and fertilizer runoff. It is unfortunate for fish that they can't just say no to drugs. And, in many ways, neither can we.

After watching Sylvia's documentary *Mission Blue*, I'm pretty sure she is more comfortable in a wet suit than a business suit. Nevertheless, I couldn't help introducing myself to Sylvia at a Moms Clean Air Force event. With a glow reminiscent of the bioluminescent fish she observes for signs of strain from the warming ocean, *Time* magazine's first Hero for the Planet leaned in to me as if she had a deliciously wicked secret and said, "Don't forget to tell your members about the oceans."

As the blue heart of our planet, the oceans inspire us to protect both salt and freshwaters. And they push me to understand how hair dye affects those waters. According to the EPA, pharmaceuticals and personal care products are contaminants of emerging concern (CECs) because they are increasingly being detected at low levels in surface water. After the chemicals in beauty products travel down the drain, they make their way through wastewater treatment plants and are then discharged into our waterways. While water treatment is important to protecting the health of people and aquatic life, only about half of the prescription drugs and other newly emerging contaminants in sewage are removed by treatment plants. The impact of most of these CECs is unclear. All fifty states and some US territories and Native American tribes have advisories about eating fish because of the toxins the fish contain.

Let's look more closely at that trip down the drain. While I could find no available data on how hair dyes break down after the chemicals enter the environment, David Lewis, a dye chemist, proposes that hair dyes can form nitrosamine compounds after disposal. The United Kingdom's Department for Business, Enterprise, and Regulatory Reform characterizes nitrosamine as toxic to more animal species than any other category of chemical carcinogen. And the Campaign for Safe Cosmetics states that nitrosamines are found in "nearly every kind of personal care product."

I work closely with organizations such as these that follow the life cycles of chemicals in personal care products. MADE SAFE is one of those organizations, founded by Amy Ziff to reduce the toxic chemicals in our lives. Amy invited me to a small conference at the Warner Babcock Institute for Green Chemistry featuring celebrated chemist

John Warner. When Amy saw that I had stopped coloring my hair and learned I was writing a book about hair dye, she said I must talk with Warner's colleague Paul Hawken because "he can go deep on the chemicals and ingredients of harm in all hair dyes. He also has facts about how these chemicals impact the hair that horrified me."

Paul Hawken is a legend in the environmental world. He's a best-selling author and an activist and has founded several successful green businesses. I had heard Paul give a charismatic talk at the Omega Center for Sustainable Living a few years back. His lecture was full of his personal brand of hypnotic energy and entrepreneurial optimism. Paul believes there's no reason we can't heal the world and rebound from climate change with creative solutions.

During a phone call with Paul, I learn that "over 95 percent of hair dyes end up in aquatic environments, causing damage to marine life and polluting our rivers, bays, and the ocean. When we use toxic chemicals, they do not disappear."

Like his 2017 *New York Times* best-seller *Drawdown*, our call contains very little small talk, just big, expansive ideas with data to back up theory.

I ask Paul, "What happens when the mixture of chemicals in hair dye goes down the drain?"

He explains that harmful synthetic chemicals—potential carcinogens and endocrine disruptors—get absorbed into our bodies during the hair dyeing process. And 3 billion pounds of hair dyes also go into our oceans and bays each year. The by-products of salons—chemicals, bleach, perm solutions, color formulas, and ammonia—get rinsed down the drain and absorbed into our waterways and into groundwater. They persist in the environment, bioaccumulating in both water and soil, affecting the tissues of living organisms, such as the fish we eat. As climate change provokes more and more extreme weather events, groundwater can become mixed with floodwaters, as happened after Hurricane Katrina. Floodwaters then carry along these pollutants, spreading chemicals that can create serious long-term health effects. We understand many of these effects but certainly not all of them.

Paul likens women to crash test dummies for the beauty industry. He believes it's time for the industry to clean up its act because "behind all the glossy ads, the long-legged models, the glamour and glitz, is a witch's brew of compounds that should not be placed on the hair, scalp, skin, or lips."

He sums up the fear that many women share whether or not they're actively investigating cosmetic safety or demanding not to be exploited. Paul warns that in a decade we'll find out that hair dye is "wreaking havoc inside women's bodies in ways that doctors do not understand, see, or correlate."

Then he tells me that, with John Warner, he's working on a hair coloring product that will be safe for both women and the environment. It couldn't come soon enough.

∽

Poet Mary Oliver asks in *Upstream*,

> Do you think there is anything not attached by its unbreakable cord to everything else? Plant your peas and your corn in the field when the moon is full, or risk failure. This has

been understood since planting began. The attention of the seed to the draw of the moon is, I suppose, measurable, like the tilt of the planet. Or, maybe not—maybe you have to add some immeasurable ingredient made of the hour, the singular field, the hand of the sower.

Environmental health advocates are the "sowers" of our time, planting important messages and holding on tight to threads of hope. In dogged determination and spirited activism, they add the "ingredient made of the hour."

I'm not so naïve as to think that ditching hair dye will save us from a regulatory system that has unraveled in the most immoral way. But, come hell or high water, and both seem more likely than ever, our health and the health of the planet will be impacted by the chemicals we pour on our heads and down our drains.

At the very least, when we stop coloring, we can take a certain satisfaction in the knowledge that we're not polluting our air or waterways for the sake of vanity. And maybe, just maybe, that small act will knit us closer to a healthier future, as this is our time. The moon is full of activism.

Chapter 7

The Road to Change

THE DECISION TO QUIT HAIR DYE is deeply personal. It reflects one's individual understanding of what is beautiful and what is not, what is age-appropriate and what is not, and what is safe and what is not. For me, attending the meeting about chemicals in Washington, DC, was the push I needed to make a change. For other women, it could be experiencing a health scare, or switching jobs, or simply being inspired by a friend.

Yet even when we experience a tipping point, change doesn't happen overnight. According to the American Psychological Association, change actually comes in five stages: pre-contemplation, contemplation, preparation, action, and maintenance. There's the trial and error, the second-guessing and chance of relapse, and the final acceptance. In between, there's the scary uncertainty of leaping into the unknown.

Even once you've made the decision to quit coloring, the hard reality is that hair takes time to grow. Transitioning to your natural color involves a substantial commitment of time, effort, and emotion. Many women start the process of going gray as if they were on a vision quest, examining their strands in the mirror, reclaiming something buried and possibly lost, probing for sparkly silver roots.

For those of us in this camp, ditching dye can seem like going on a diet. We become obsessed—not with food but with how fast our hair grows. The days between the last dye application and a full head of silver hair creep by, painful as the growl of an empty stomach.

For others, the widening skunk stripe is too much to bear. Rather than going cold turkey, maybe they opt for a pixie cut or blend incoming silver strands with dyed highlights. Whether it's losing weight or quitting hair dye, there is no single solution that works for everyone, and every path involves soul-searching. But regardless of the method you choose, there's no getting around the fact that going gray is a waiting game.

Given that this process requires patience and mental stamina, it's worth remembering that coloring is no piece of cake, either. So far, I've focused on my third reason for quitting hair dye: the chemicals. Let's take a minute to examine the other two: the upkeep and the cost.

The Upkeep

Taking care of dyed hair is far from cut and dry. Not every person requires the same amount of maintenance—the upkeep depends on how fast your garden grows.

Average hair growth is one-half inch per month, but metabolism, diet, and genetics all affect the speed. Different growth rates mean the time between root touch-ups or full-head colorings is distinctly individual.

If you're dyeing gray regrowth a dark shade, the demarcation line between natural and dyed hair is stark, and you could need color touch-ups every two to four weeks. For those with naturally light hair, gray is not as noticeable. Of course, if you hate seeing any gray, the time between dyeings is as soon as you can't bear the sight of it any longer.

The amount of time required for coloring also varies, but it averages about two hours. This includes application, processing time, rinsing, and drying. Of course, if you include a cut, you'll be in the chair longer.

I have almost one inch of new growth per month, which meant I needed to color my part every two to three weeks. I used to allot three hours every month for a color, cut, and blow-dry.

The Cost

The cost of hair dyeing also varies greatly. Not surprisingly, the major difference comes down to whether you buy a color kit at the drugstore or pony up for a professional salon visit. At-home hair coloring offers an inexpensive DIY approach, while professional colorists can use their broad experience to mix a few shades so the hair has a natural-looking depth. That is much harder to achieve with out-of-the-box color.

If you go the salon route, the total cost will depend on the amount of time between colorings, what you're having done, who you're seeing, and where the salon is located. Prices also can vary from stylist to stylist within a salon.

The average woman dyes her hair every eight weeks, with 37–52 percent of women coloring at home and 48 percent going to salon professionals. So the price differential is huge. At Walmart, a box of hair dye costs $2.99–$19.99. According to *American Salon* magazine, the average cost of a hair color service at a salon is approximately $85. But a salon visit in a major city can cost upward of $500.

I dyed my full head every four to six weeks and needed to dye my part every two weeks. I went to a local salon for the full shebang and purchased a "natural" permanent hair dye, Naturtint, on Amazon or at Whole Foods, for at-home root touch-ups. This cost about $15 per box. If I had a special occasion on the horizon, I'd spring for a professional root touch-up and blow-dry, since the box color didn't cover as well or last as long. In fact, about ten years ago my family went to Spain to visit relatives for the better part of a month, and I knew I would need to apply color to my part at least once. I worried during the whole plane ride about the dye exploding in my suitcase. As it turned out, I needed to apply the color twice, in a bathroom the size of a kid's closet.

I live two hours north of New York City, and the price of my professional hair color (with Aveda products) was around $125. A salon root touch-up, or, as my former hairdresser would call it, a 911 appointment, was between $40 and $50. For some reason, the price was different depending upon who was at the reception desk. These salon prices did not include cut, blow-dry, and tip. The full cost of my last hair coloring visit clocked in around $250.

Radically driving down my hair budget and gaining precious hours were fringe benefits I hadn't conceived of when I quit dyeing. I now get haircuts every two months at $60 per cut, plus a 20 percent tip. I allot an hour for the cut and blow-dry. It costs me $432 per year, as opposed to $3,000 per year when my hair was professionally colored. As for the time saved each year that I don't spend coloring, I get back thirty hours of my life.

Of note, globally, the most expensive place to get a haircut is Switzerland. No matter where you live in the world, women pay about 40 percent more than men . . . with the exception of France's former president François Hollande. It was reported that he spends more than $10,000 per month to maintain his locks.

Making a lifestyle change is a process that takes time and requires ongoing support. My husband was extremely supportive. But since he had never dyed his hair, he couldn't really relate to the beauty angst women experience when they grow out long-dyed hair. I found that I had to approach the topic of chemicals with my girlfriends very gingerly, as any mention of the toxic overload from hair dyeing could easily devolve into a series of not-in-my-backyard-style proclamations: "It's a good look for you, but it's not for me. Hair color gives my hair life." My sense was just the opposite. The chemicals felt like a death sentence to me. Some friends said they would consider going gray one day. One friend told me repeatedly during my transition, "You can go back to dyeing anytime you want if you don't like it, you know." I must confess that I held on to all my boxed dye and hair coloring paraphernalia, just as I hold on to my larger-size clothes when I lose weight. Just in case.

The Going Gray Crowd

I found a tremendous amount of support online. Surfing my way to gray on Facebook, Pinterest, and Instagram offered me endless photos of silver hair. These images spanned a blended spectrum of gray and bolstered my determination. Whenever I passed a mirror, my fringe of opalescent silver hair caused me to do a double take. But when I looked online at women with their skunk roots, rumpled ombres, bluish tints, multitonal tresses, and pearly white sheens of various lengths, I found answers. This sorority of silver sisters revealed their before-and-after transition photos on private Facebook pages titled "Going Gorgeously Gray," "Gray and Proud," "Going Gray, Looking Great," and others. So many of the photos were inspiring. Others were startling: shaved heads, corrective interim dye blending jobs, purple- and orange-tinged hair, dried-out tresses, and pixie after white-hot pixie. Each was empowering in its own right. Strangers in the throes of a transformation were struggling with the gargantuan beauty task of living with daily hair chaos while staking out a new set of rules for beauty and aging.

Despite some misgivings, these women were creating new senses of self, exhilarated by the luxurious outpouring of hope, faith, and honest longings. Some confessed to being unhinged by the reality of slow hair growth and shifting cultural beliefs . . . and their own reflections in the mirror. Once they overcame this hurdle, the women seemed to discover a feeling of rightness and calm acceptance. Their contagious musings continually kept the going gray community afloat.

These growing social media pages inspired my "secret" Pinterest board that contained images of silver-haired beauties and hairstyles I coveted. Privately, I loaded up the board with stunning silver cheerleaders pushing the limits of going gray. At the time, the sharpest silver pins in the bunch were a handful of gorgeous models rah-rahing the cause.

The late Cindy Joseph's attitude about aging was what allowed her to go silver gracefully. Silver hair was her medal of honor, one she felt she'd earned by living a rich and passionate life. Cindy described how she had felt trapped before she quit dyeing her hair in her late forties. She felt her energy was stuck because she was hiding something. When she started to go silver, Cindy defined it for herself and others: "I use that word because silver is valued in our society." She had a distinctive silver streak by age thirty-eight, and by forty-three she was silver all around her face. In those early years, Cindy was not comfortable with her silver, but after a few years of dyeing she realized she was covering up what she was celebrating—her age. She threw out the bottle of dye and transitioned to silver. The day Cindy chopped off the last bit of dye, she was approached on the street by a casting agent to model for the fashion brand Dolce & Gabanna. An unexpected modeling career took off, landing her a contract with Ford Models. Sadly, Cindy recently passed away in her mid-sixties.

Yazemeenah Rossi's loose, voluminous waist-length natural hair is seriously breathtaking. She's never dyed it. Yazemeenah is the silver-haired "It" model for TV and print ads, often portraying older women. Born in Corsica and now living on the coast of California, she's a free spirit who doesn't want to spend hours dyeing her hair. "I'm passionate about what I'm doing. I have no time." The same age as me, Yazemeenah has a seemingly eternal beauty that flourishes inside and out.

Danish actress and gray-haired J. Crew model Pia Grønning claims she often gets modeling requests from hair product manufacturers. Without apology, she says, "I don't take them because I know they'll dye my hair. And they always want to cut it too. Like, why? Can't a 65-year-old have long hair?" Pia was originally blonde, which made her natural transition look effortless. Sigh.

What these women have in common is that they went gray, kept their hair long, and strived to live a healthy lifestyle. From these role models, I found encouragement to reach my ultimate goal: long, silver healthy hair. With the lack of color, I confess that I do not feel I look young anymore, but I also feel somewhat ageless. And that gave me the confidence to finally throw out a bathroom drawer full of at-home hair dyeing paraphernalia.

Hair Bias

I found my own hair transition tough enough. But women of color face a whole different level of hair scrutiny. To see how beauty standards affect African American women, the Perception Institute conducted a "Good Hair" study in which more than 4,000 people took the Hair Implicit Association Test. The test measured unconscious negative judgments about natural, textured hair. The majority of test participants, regardless of race, showed bias against black women's natural hair. Perhaps these prejudices against black hairstyles should not be surprising. Remember when in 2016 a US circuit court ruled to affirm a company's right not to hire applicants with dreadlocks?

The "Good Hair" study also explored women's feelings about their own hair. It found that most women worried about their hair, but black women experienced higher levels of hair anxiety than white women. When it came to the workplace, one in five black women felt the social pressure to straighten their hair for work—twice as many as white women.

As we've seen, chemical straighteners have real implications for women's health—as do other products specifically targeted at women of color. A colleague of mine, Shakeila Stuckey-James, told me about a study that immediately caught my hair-obsessed attention. Conducted by Black Women for Wellness, the study was the first of its kind to examine the concentration of endocrine disrupters in hair products marketed to black women. Marissa Chan, the organization's environmental research and policy coordinator, summed up the findings: "Black women are the most over exposed and under protected to toxic chemicals in personal care products. Black hairstylists are some of the largest impacted." When a reporter asked what the FDA was doing to protect black women, Chan replied, "FDA is aware of the chemicals, but they don't have the power to regulate." But according to Chan, the issue runs deeper than simply policy. "Black hair is political. Black women feel the need to straighten their hair." To confront issues of social justice, Black Women for Wellness has created a Healthy Hair Initiative.

There's no splitting hairs over the bias faced by women of color. With perceptions about hair affecting everything from self-esteem to job retention, the pressure to conform to potentially unhealthy beauty standards is real.

The Long and Short of the Workplace

I could feel my attitude shift as my hair tone went from inky black to dove gray. I cut the dyed ends off every few weeks, with a big chop when my daughter, Lainey, got married. While the dyed ends were definitely not "good hair," I felt that with shorter, grayer hair, I looked older.

I notice that baby boomer women, those who have grown up expressing themselves by bucking hair rules, gain confidence from letting their hair do what it wants to do. Yet they also face a paradox. As Rose Weitz, author of *Rapunzel's Daughters* points out, "Short hair in contemporary American culture is typically seen as less sexy, but more professional. Women are expected to be feminine, but also are expected to fit in with men's norms in the workplace, in which, more often than not, they're working with male bosses and working with male higher-ups, so, that's always a trade-off."

Joan Juliet Buck, who was the American editor in chief of *Vogue Paris*, takes a different view. She explains the appeal of short hair: "Short hair removes obvious femininity and replaces it with style. . . . Short hair makes other people think you have good bones, determination and an agenda."

At a lunch meeting with an executive I hadn't seen in a while, a woman I admired for her intellect and steadfast kindness, I was greeted with a personal question: "You have such a young face; why would you do that to your pretty appearance?" After the wedding, I had let my hair grow a few inches past my shoulders.

Sputtering out my going gray mantra, perhaps a little too quickly—"Oh, I got tired of the upkeep, the . . ."—I pulled back, realizing that she was commenting not only on the color of my locks but also on the fact that my hair was gray *and* long.

Luckily, I work mostly at home, so letting my hair go gray and grow long with a blunt cut, without blending, worked best for me. It fit my lifestyle and fashion sense. I quickly discovered that not everyone can take this route.

This is where the online support can be extremely helpful and forgiving, especially for those who recalibrate their decision as they transition and decide they need blending. A few even go back to dyeing, with little judgment from the going gray crowd.

Many women cite gray hair as a job liability—a kind of sell-by date for the workplace. Some found their jobs expired or were in jeopardy when they went gray. Even leaving hair aside, aging is hard on women in the workplace. While going gray can set an empowering model of age diversity, I know personally that being the oldest, grayest woman in a conference room filled with younger colleagues can sometimes feel diminishing. Plus, unemployed older workers who get rehired face earnings losses of 25 percent compared with the income from their previous job. We know sex discrimination in the workplace is common, but the combination of sexism and ageism puts older women at a distinct disadvantage.

Going gray is never an easy road, and if you dyed for as many years as I did, it can be a long one. But it shouldn't have to be lonely. Whether you're worried about how your hair will affect your job or are simply fed up with skunk lines and brittle dyed ends, know that there are many other women facing similar struggles. Each woman has to pave her own way, but you can still share the journey.

Chapter 8

A Greener Shade of Gray

MAYBE YOU'VE GOTTEN THIS FAR: you've read the warning on your boxed dye and checked the hair color company at the Environmental Working Group's Skin Deep website. Or maybe you've asked your colorist to please keep her gloves on and open the salon window. Or you may have decided that a cold turkey transition to gray is just not for you and you'd like to blend in some "natural-looking" highlights, but you're still worried that conventional hair color is harmful. You may ask, Why can't they just make a safer, less toxic alternative to traditional hair dye?

For years, I stuck my dyed head in the sand, unable to face my own body pollution quandary. So I get it. I realize many women won't stop coloring unless their health depends on it—pregnancy, allergies, cancer. And even once you vow to give up the dye, a million little insecurities can send you back to the salon sink to blend a skunk line or add some highlights. Even though my personal environmental ethics kicked in after that fateful meeting with the scientist in DC, I'm still not sure I would have transitioned cold turkey if I could have eased into the gray scene with safer hair color. I've asked myself more than a few times, *Would I have kept coloring if there had been a totally nontoxic alternative?*

As the culture shifts, easing up on gray hair rules, change is trickling down to the salon business. There's news on the horizon about safer alternatives for those who want to help their gray hair along or who just can't stand the sight of gray. And with gray hair becoming more and more popular, the hair color business is trying to keep pace. Some hair color manufacturers are taking notice and attempting to adapt. Cyrus Bulsara, president of beauty data specialist Professional Consultants and Resources, says that at-home mass retail hair color offerings generate $1.4 billion in annual revenues. The salon hair color business generates $901 million at manufacturers' prices. It's estimated that covering gray is responsible for half of hair color sales. Bulsara says hair color services are slowing down. He thinks the dip is "due to high costs and lower salon visit frequencies, plus a growing number of women embracing their natural gray, silver or white." It may also be due to women making more informed and less toxic beauty choices.

If we had better regulations and health protections, we'd have safer chemical dyes. But in the absence of government reform, some salons are responding to market forces.

According to an article from Beauty Independent, salons and hair care brands are divided on what to do with the going gray crowd. For colorists willing to ride the gray wave, "gray blending" services provide a new revenue stream.

The article cites Maida Salon owner Farah Hurdle, who blends her own gray hair with highlights and provides a blending service (highlights and lowlights) to help women transition. She has also noticed more of her younger clients deciding not to cover up their gray. Once Farah started offering the blending option, new clients flocked to her salon. Even though they book fewer appointments (because of less permanent hair coloring procedures), Farah doesn't worry about financial loss. "I am loving this transition in the beauty industry. I am here to help people feel better and look better too."

While this may be a hard sell for some in the hair business, others are trying to stay a step ahead of the going gray curve. Ann Kohatsu, director of product development at Hush, a company that sells a temporary hair color spray, says, "The gray shade of this product will attract the women that have embraced their natural grays and have the desire to disguise sparse areas." Knowing the gray-hair trend is here and most likely here to stay, Kohatsu expects Hush's sales of the gray shade will increase in the upcoming years.

What about safer dyes? There have been recent break-throughs in permanent and semipermanent hair dye, as natural beauty takes hold of lifestyle trends. Since I started writing this book, a number of safer formulas have been developed.

According to colorist Stephanie Brown, there's a big difference between conventional dyes and organic hair dyes, which contain fewer chemicals and usually no ammonia. Ammonia swells the hair cuticle so that the color can penetrate. It's got an unpleasant odor, can cause allergies, and can also weaken and damage the hair, leaving it dry and brittle. Stephanie says not all organic hair color is chemical-free. "While chemicals are present, they rely on botanical ingredients and naturally-derived ingredients. The only truly 100% natural hair dye is henna."

Henna, made from plants, coats the hair without stripping or penetrating the hair shaft like chemical hair dyes. It's part of a long tradition of dyes made from leaves or other parts of plants that have been powdered and dried. The earliest use of plant dyes dates back to the Paleolithic period, when natural dyes were used to adorn dwellings, clothing, and bodies.

I remember when henna was trendy in the 1970s. Henna turned hair an unnatural color similar to the paint job on my first car, a 1975 burnt orange Datsun B210. With women now opting for natural living and DIY lifestyles, henna is making a comeback. Hand-blended powders that include hibiscus, turmeric, saffron, beetroot, and coffee are replacing chemical plumes with appealing smells. Most henna is 100 percent natural, vegan, and preservative- and synthetic-free. Because henna doesn't alter the hair's natural structure, pregnant women and cancer patients who have been cautioned against using chemical dyes may find it a viable alternative.

Henna generally lasts four to six weeks, depending upon how frequently hair is washed. It fades as the hair grows, unlike permanent chemical hair dyes. Unfortunately, even henna products can have a downside. Stephanie warns that henna-based dyes are "actually pretty harsh on the hair because they contain metallic salts." Some also include chemicals such as PPD and other additives, so it's important to check the ingredients.

Rachel Sarnoff, former chief executive officer of Healthy Child Healthy World, with a popular TEDx Talk, "Can One Straw Change the World?," has beautiful long brunette hair that she's dyed using henna. Rachel had tried henna in college when she was experimenting with less toxic natural hair color to dye her hair red. But then she wanted to cover her gray. Henna can be messy, so she asked a hairdresser friend to apply the color.

Rachel describes the benefits she discovered when using henna:

- My hair is super shiny.
- It feels thicker and deep conditioned.
- The henna colors my grays so that they look like tiny, very fine highlights.
- The color lasts for two months and fades out gradually, without changing the base color of my hair.
- I'm not putting any toxic chemicals onto my body and down the drain.

Rachel compares henna to what hair coloring used to be like "before there were thousands of shades of toxic chemicals calibrated to deliver a very specific hue." She likens the difference between hair dye and henna to the difference between painted and stained wood. "A light colored wood will look the same as a dark colored wood when painted; but a light colored wood will take stain completely differently than dark."

Even the corporate hair world is now taking a page from the henna playbook. Hair giant L'Oréal is poised to introduce the first vegan, 100 percent plant-based hair dye collection, called Botanéa. According to a recent *Forbes* article, Botanéa will be available in European salons soon. L'Oréal set its sights on women who want transparency and cleaner products and to lighten their toxic loads, stating, "Professional hair care has taken a hit since it's traditionally been difficult to formulate professional hair dye without ammonia and other potentially dangerous chemicals." Botanéa is sourced from three plants found in India.

This plant-based formula sounded promising to me, but I couldn't help remembering that even some plants are toxic. After my long history with hair dye, I maintain a healthy skepticism about any product claiming to be 100 percent safe and natural. After all, I used Aveda hair dye for years, thinking the company's mission was as pure as the driven snow. Statements like this one from the Aveda website sound convincing: "We connect our Mission with product development by using what we define to be green ingredients whenever possible."

The key phrase here is "whenever possible." What about when it's not possible? And should we trust Aveda, rather than medical science, to "define" what's green?

Aveda also touts that its "products are formulated without parabens, phthalates and sodium lauryl sulfate."

This is good, although the company neglects to mention what it's substituted for those ingredients. As mentioned earlier, some ingredients typically substituted for SLS may be just as harmful as the original. The Environmental Working Group identified twelve possible surfactant cleansing agents that could be swapped out for SLS.

More from the Aveda site: "We work hard to ensure that ecological and cultural diversity is represented by responsibly sourcing key ingredients from different habitats all over the world."

Key ingredients are not all of the ingredients. I suspect there may be more than a few that fail to live up to these high-minded ideals.

The company's definition of a green ingredient requires that it meet at least one of the following criteria:

- Naturally derived, which we define to be those for which more than 50% of the molecule comes from a plant, non-petroleum mineral, water, or some other natural source.
- Certified organic.
- Sourced from sustainable or renewable plant-based origins, and does not negatively impact the ecosystem.

This all seems nice, but one criterion is not enough. And Aveda never claims that its products are 100 percent certified organic.

To me, Aveda embodies the mix of idealism, compromise, and drive for profits that so often defines "natural" companies. The brand has an interesting backstory, though. It was founded in the 1970s by Horst Rechelbacher, a German hairstylist. His mother was an herbalist, and Horst traveled in the Himalayan region of India, sourcing the most holistic ingredients he could find. With a passion to develop healthy hair products, Horst even made Aveda's packaging admirably sustainable, using 100 percent post-consumer recycled PET (polyethylene terephthalate—a plastic). Aveda was sold to the Estée Lauder Companies for $3 million in 1997.

Before his death in 2014, Horst set out on a mission to inform the public that most beauty products, even "natural" or "organic" ones, contain many harmful petroleum-based ingredients. Once his noncompete clause with Estée Lauder expired, Horst put a large amount of his fortune behind a beauty line called Intelligent Nutrients. He grew most of the ingredients used in the products on his organic farm, citing the health of salon workers as one of his primary motivations. "All my old colleagues are dead of bladder, liver and lung cancers. Hairdressers don't live long lives." Horst died at seventy-two of pancreatic cancer.

I asked the author of the blog *This Organic Girl*, Lisa Fennessy, about her thoughts on organic hair color. For an article titled "Organic Hair Dyes: The Good. The Bad. The Ugly.," Lisa contacted ten "natural" hair color companies to ask for their ingredient lists. She identified PPD, coal tar, and SLS, or derivatives of these chemicals, as toxic culprits in many of the organic hair dyes. The healthy-sounding list included Simply Organic: Organic Way or Oway; Naturcolor, Madison Reed, Organic Color Systems, and Clairol's Colorblend. One even employs a "farm to chair" slogan.

Lisa noted, "The one point that I really want to drive home here is when a hair dye is labeled 'organic' or claims to be 'natural,' don't be fooled into thinking you are getting a healthy alternative." Lisa said she was hard-pressed to find a truly chemical-free hair dye that both covered well and looked good. She concluded that her two healthiest options were henna and Hairprint.

This wasn't the first time I'd heard of Hairprint. In fact, it's the product Paul Hawken mentioned when we spoke. Paul and his wife, Jasmine Scalesciani, Hairprint's cofounder, worked with chemist John Warner to create a natural way to cover gray. Dr. Warner's hair had gone gray when he was a student at Princeton. Experimenting with plants and oils, taking his cues from nature, he was able to restore his hair to the same hair color he'd had as a young man.

Hairprint's complete list of ingredients are all rated #1 on the Environmental Working Group's Skin Deep website, the lowest numeric hazard ranking (#1 is "low concern"; #10 is "high concern").

"It's not a hair dye," Paul said before I could ask the obvious question.

Sensing my skepticism, Paul repeated his initial comment a few times and added, "Hairprint works with the hair's natural structure to recreate the color it was before it turned gray. It doesn't strip the hair like traditional hair dye." It's called Hairprint because "like your finger with its fingerprint, every strand of hair on our head has a unique identifier associated with its original color, even if it has turned completely gray."

Thank goodness he couldn't see me scratching my head and thinking he'd drunk the Kool-Aid. But honestly, I stayed open to his explanation because I could tell that describing Hairprint was tricky. And I already knew Dr. Warner from my visit to his institute in Massachusetts with my colleague Amy Ziff. Warner is considered the founder of green chemistry, a discipline that employs benign, nontoxic molecules to achieve better results than synthetic chemicals.

Paul continued, "Hairprint doesn't deposit color onto or into your hair, like conventional dyes; it creates a nontoxic chemical reaction that changes the strand pigmentation, re-creating the hair color you were born with."

Obviously, I haven't tried Hairprint because I was already in full transition mode when I learned about it. Since I was not planning to put more products back into my personal plume, I read testimonials and a few product reviews instead. One interview with Paul mentioned that he took "a spoonful of the black stuff" and began eating Hairprint to show how safe it was. A testimonial by Kristen Arnett of Green Beauty Team summed up the hair product this way: "All in all, Hairprint True Color Restorer is an impressive product. Its ingredients and the way it functions are revolutionary and unparalleled. Particularly if you are in the brown-haired category and are simply looking to turn your grays back to your normal color, this is a fabulous non-toxic option."

While Hairprint doesn't work for blondes, redheads, or those who have more than 50 percent gray, Paul is convinced that once the product is perfected, the company will be able to expand its color palette.

Back in the lab, scientists continue to concoct new chemicals to dye hair. At the same time, they are searching for ways to make dyeing safer and less harsh. Researchers at Northwestern University think they may have cracked the code by using graphene. Graphene is a material composed of a single layer of carbon atoms in a honeycomb formation that's used in electronics and other devices. It's nontoxic, chemically inert, and naturally black in color. In a study published in the journal *Chem*, Dr. Jiaxing Huang and his team recounted how they prepared graphene oxide with a gel and then sprayed it on platinum blonde hair. It dried in ten minutes and left the hair coated a dark color, creating a natural-looking hair shade. It even stayed on the hair after thirty washes. The researchers are working to perfect a method that could make graphene hair color a safer alternative to conventional hair dyes.

Dr. Huang commented, "Because we now have a coating-based dye, we don't have to get into the hair or change the chemical structure. It's a nanomaterial solution to solve a chemistry problem." That chemistry problem is the toxicity of certain chemicals in traditional hair dyes that serve up a bevy of health issues, both known and unknown.

Will graphene be a spray-on miracle for the hair dye industry? Not all scientists think so. Especially those who study the potential health and environmental impacts of engineered nanomaterials.

Andrew Maynard, director of the Risk Innovation Lab at Arizona State University, explains, "Engineered nanomaterials like graphene and graphene oxide (the particular form used in the dye experiments) aren't necessarily harmful. But nanomaterials can behave in unusual ways that depend on particle size, shape, chemistry, and application. Because of this, researchers have long been cautious about giving them a clean bill of health without first testing them extensively." So far the research doesn't indicate graphene is dangerous, but more study is needed.

If graphene sounds too high-tech for your taste, there's an alternative you *can* actually taste. Promising studies by the American Chemical Society report that scientists have developed what they consider a truly natural, nontoxic hair dye derived from black currants.

The researchers extracted and purified pigments from black currant skins called anthocyanins, which produce colors ranging from pink to violet in fruits, vegetables, and flowers. (Anthocyanins are typically used to make natural food dyes.) The pigments were then made into a dye paste and applied to bleached human hair. The hair turned bright blue and, remarkably, stayed that way for twelve shampoos. Fortunately, the researchers think that with modifications to the formula, there will be more options beyond the Marge Simpson look. Still, black currant dye may be best suited to the bold, with red and violet colors.

Funky fruit-based hair color may the wave of the future, but we're not there yet. And even if we manage to start washing nontoxic dyes down the drain, can salons really become sustainable? Every day, hair salons across North America accumulate over 400,000 pounds of waste, including hair clippings, foils, paper packaging, and plastic tubes. Given all the materials that go into hair styling, is there such a thing as an eco-salon?

Well, it just so happens that progressive business owners can apply to be certified as a Green Circle Salon. Started in Canada in 2014, Green Circle Salons collect, recycle, and repurpose hair clippings, used foils, color tubes, excess hair color, papers and plastics, glass, and other waste and divert them from landfills and waterways. As much as 95 percent of the materials that would be thrown out can be recovered. The salons use less dye while also saving energy and water.

CBS Minnesota highlighted the state's first green salon, Wave Salon. In the back of the shop, there's an extra bag of hair dye waiting to be recycled that includes "all chemicals that in the past we would have dumped it down the drain. And it would have gone into our water systems," said Wave Salon's co-owner.

Foil used in salons typically goes into landfills, where it will stay until it finally breaks down—four hundred years later. While only 1 percent of salons recycle foil because it's difficult to clean off the dye, Green Circle Salons are able to separate the materials, allowing foil to be recycled. Even the cut hair is sent to a women's correctional facility, where it is turned into recycled "hair booms" that can be used to help clean up oil spills.

While all these efforts might seem expensive, it's pretty easy to offset the costs. At Wave Salon, customers pay a $2 "Sustainable Stewardship Fee," which seems like a small price for knowing that you're helping to protect the planet. Clients interested in voting with their pocketbooks can support salons that support healthy communities. Seems like a win-win to me.

Green Circle Salons provides an online green directory that makes it easy to find participating salons. I put my address into the directory and ten salons came up within one hundred miles.

Even if there are no Green Circle Salons near you or you prefer to dye your hair at home, you can adopt some of the salons' sustainable practices. Rather than pouring waste dye down the drain, consolidate all the liquid into one bottle. That way, only one bottled chemical cocktail is making its way to the local landfill.

If you're not ready to break up with your hair color, flirting with less toxic henna or Hairprint might be worth a whirl. I'm still not crazy about the idea of eating it, but I admire the scientists who are trying to improve on the color wheel of nature, without destroying nature in the process. Or, if you're ready for something truly wild, ditch the dye altogether—go cold turkey. Rather than looking for color in a bottle, find it out in the world, in those anthocyanins that give plants and flowers their vibrant hues. It's always a good idea to take a hike!

Chapter 9

Polishing the Silver

S PRING UNFURLS SLOWLY ON MARTHA'S VINEYARD, where I've been spending the winter and spring writing. The weather turns us from a cold beach walk to a hike on a protected inland Nature Conservancy trail. Along the wooded path, my friend Ali and I are deep in conversation about mothering, activism, and life.

As we meander, we stop often, reminding ourselves to pay attention, to absorb the wild beauty of bright greenery that rises up through sandy trails marked by bluer-than-blue ponds, expansive fields, and an unapologetic over-growth of dense wilderness. In amazement, we stop along the trail to notice how much living things yearn to spring up. Despite the uprising, the landscape is still suffer-ing from nighttime snaps of biting temperatures, as the weather in these parts hangs on to the last threads of cold. An island surrounded by water allows the ocean to always have a chilling last laugh. Next week, the temperature is predicted to be a welcomed twenty degrees warmer. But my friend Ali, a full-time Islander, reminds me that the weather here, with its unique microclimates and variations in temperature, is too unpredictable for us to rejoice just yet.

"Everyone deserves clean, healthy food," she asserts. "Everyone deserves safe, healthy products," I reply. Ali and I are both activists. As a trained teacher, activating parents and children feels like second nature to me. Ali shares this affinity as an author, NPR radio host, and founder of a grassroots group that encourages children and their families to eat healthy, locally grown food. We work in different fields, but, like this particular landscape, where deep forests meet open meadows, our paths naturally converge to confront the overarching crisis of climate change. With our children grown, we feel hardy and strong, but we carry the weight of leaving behind a healthy, just, and equitable world.

Roaming through the meadow, I catch sight of a periwinkle-winged flurry, like a gossamer cloud. As Alice Walker writes in *The Color Purple*, "I think it pisses God off if you walk by the color purple in a field somewhere and don't notice it."

Butterflies so soon? Yes, we marvel. Stopping to identify a tiny purple butterfly with twinkling pink-pewter undertones using my phone, I learn these "azure" butterflies may have peeled out new wings overnight like flowering petals, as they appear for just a short time in spring. The sublime weightless creatures also caught the attention of New Englander Robert Frost:

Blue-Butterfly Day
It is blue-butterfly day here in spring,
And with these sky-flakes down in flurry on flurry

There is more unmixed color on the wing
Than flowers will show for days unless they hurry.
But these are flowers that fly and all but sing:
And now from having ridden out desire
They lie closed over in the wind and cling
Where wheels have freshly sliced the April mire.

Not surprising that the great poet of snow would compare one-inch butterflies to *sky-flakes*. The tiny purple spring trailblazers, *flowers that fly*, catch my friend and I in the same thinking of another, more contemporary poet: Prince.

It was rumored that Prince loved the color purple because it was considered the dramatic color of royalty. Purple is energetic—*flurry on flurry*. Purple is Prince, who had died the night before at age fifty-seven. We mourned. How could it be?

I have a hard time believing that I believe in signs—that the universe is trying to tell me something. But I do believe in serendipity. The word plucks an emotional chord. In an open field on a small island, a coincidence of color feels potent. So, after the hike, I allow myself to muse, wondering if connecting Prince's death to purple butterflies will be the nudge I need to finally quit my last chemical hair crutch: purple shampoo. The shampoo is marketed to women with gray and platinum hair to neutralize yellow tones. But it is not without pitfalls. What pisses me off about purple shampoo is the pungent smell disguising a riotous cocktail of chemicals.

I count backward three years, to when I was fifty-seven and a full-on brunette. Now sixty, I've transitioned to long silver hair. Like many women with hair that is silver, salt-and-pepper, gunmetal, pewter, steel, iron, chrome, gray, or chinchilla (yes, I've heard that description), I find I need different products to keep my hair soft, bright, and manageable than I did when it was dyed a deep dark brown. Now, a bevy of enemies lurk in everyday life that threaten to sweep my uncolored locks into a tangle of bad hair days. Hair picks up smoke and other pollutants in the air. Chemical processes such as perms and hair relaxers, as well as hair dryers, curling irons, and straightening irons, all make hair more porous and prone to discoloration. Certain oils, emollients, preservatives, and colors in hair products deposit a yellow cast on shafts of hair. Days at the beach and backyard pool can also wreak havoc on silver hair. Ultraviolet rays from sunlight make hair more porous, and salty ocean water dehydrates hair, while chlorine can deposit a greenish tint.

Then there's the mother of all adversaries for the silver-haired woman: hard water. It may seem the water that flows from the faucet is natural and pure, but it's filled with minerals and chemicals, some natural and some added during water treatment processes. Hard water can be high in iron, like mine, and it can give hair an orange, red, or yellow cast. The pipes that bring water into the house can be filled with nitrates, chlorine, and whatever else water picks up on the joyride up to the showerhead. All of it adds to the gunk that builds up on hair from hard water.

When I first started looking for ways to counteract these effects, my research kept pointing to the much-touted purple shampoo. A far cry from the pastel color of spring butterflies, the shampoo's concentrated tint glistens like fluorescent grape gumballs. I was leery. But violet-tinged shampoo formulas are said to neutralize yellow and also help brighten silver hair. On the flip side, I read that purple shampoos are intensely drying and, if used too often, conjure up "blue-haired old lady" syndrome.

Many moons ago, hairdressers treated women's gray hair with a blue rinse, in a manner similar to laundry bluing, using a very fine blue iron powder dye. With any luck, this would tone down the yellow and leave hair a silvery white. They quickly found that not all shafts of hair absorbed the same amount of dye, leaving many older women with distressing hues of blue hair.

Today's purple shampoos promise to bring out the silvery brightness in gray hair while neutralizing brassiness, intensifying everything natural. Because yellow and purple are opposites on the color wheel, the purple pigment counteracts the warm tones.

Step into My Shower

My garden of natural roots is nourished by hard, crusty water. In a bathroom that my husband lovingly designed for me, the shower water pressure is as light as a feather and the water is as hard as the river rocks he laid, one by one by hand, on the shower floor. Turn the pressure on high, and out sputters a pathetic trickle. Water saving? Yes. Less minerals? No. The mineral build-up on the showerhead is massive, even after I give it an overnight vinegar bath. But low flow is not the only bane of hard water. The hardness factor interferes with almost every cleaning task, turning whites the color of tangy creamsicles, leaving a thin, gritty film on wine glasses, and tarnishing my super-porous virgin hair a dull, drab yellowy tone.

Squinting in the shower to read the smallest printed directions I had ever seen, or not seen in this case, I lathered up with one of the purple shampoos suggested in an *InStyle* article. The label simply read, "Shampoo. Rinse. Repeat." I marveled for a few minutes at the bright blue plumy water cascading down my body. Emerging from the shower to check the results, I saw that my roots were shiny, for sure, and that my hair had turned a beautiful tribal blue.

Grabbing my white robe, I reread a part of the article that I had skimmed over. A master stylist suggests not leaving purple shampoo on the hair too long, "as the hair could grab the purple tones!" If the stains on my robe were any indication, I had possibly overdosed on purple shampoo.

Believe me, my favorite colors are deep purples, the color of eggplants and dark-pigmented blues. In fact, I love shibori, an ancient Japanese free-form dyeing technique used on fabric, a chic progenitor of tie-dye. But shibori was not in my jeweled hair plan. How long *had* I been in the shower? Had I now joined those blue-haired ladies of yesteryear? With magnifying glass in one hand, I reached for the bottle with the other to read the fine print. I zeroed in on one ingredient, Violet #2. The Environmental Working Group's Skin Deep website stated, "Violet 2 is a synthetic dye produced from petroleum or coal tar sources." The EWG goes on to say that the dye is FDA-approved for use in pharmaceuticals and cosmetics.

Purple shampoos have coal tar? Ugh. I felt as if I'd jumped into a Looney Tunes cartoon, the one where the character gets thrown into a dark, cavernous vat of coal. To make matters worse, I knew about the threats posed by coal tar (and its derivatives), a known carcinogen found in many shampoos, soaps, hair dyes, lotions, and tattoos. I also knew products with coal tar have been pulled from the shelves in many European countries. I had ditched hair dye because of exposure to unwanted chemicals, but once again, I was choosing hair over health.

I'll confess that dyes, particularly blue and violet dyes, were already on my radar. But while I thought I had scraped my way out of that dark vat, my getaway wasn't fast enough.

Only Fools Dye Their Young

I have been reading food labels incessantly since Lainey was eight years old. It started with a breakfast cereal that made grandiose claims: "All Natural Berry, Berry Goodness," "Kid Approved," and "Contains Healthy Antioxidants." After ingesting bowlfuls of her new favorite cereal, my young daughter started to display frightening symptoms. First, she developed a headache. So we gave her Children's Tylenol. The headache got better. Then she broke out in hives. We gave her Children's Benadryl. Very quickly after taking the antihistamine, she complained her throat felt weird, like she couldn't swallow. We rushed her to an allergist, who confirmed what we suspected. My daughter was allergic to a food dye. FD&C Blue No. 2, a common food dye, was an ingredient in the cereal and in the over-the-counter children's medicines I used to treat her.

A 2003 study published in the journal *Food and Chemical Toxicology* found that blue dye used in edible products might be doing more damage to our bodies than was originally thought. A research team from Slovak University of Technology studied two blue dyes: Patent Blue and

Brilliant Blue. Patent Blue is banned from food products in the United States, but Brilliant Blue, also known as FD&C Blue No. 1, is used in food, textiles, leathers, and cosmetics. Both dyes were linked to attention deficit hyperactivity disorder (ADHD), allergies, and asthma. When Brilliant Blue dye was used in feeding tubes, the FDA issued a public health warning. The advisory noted that the side effects included blue-tinged skin, urine, and feces, as well as hypotension and death.

The scientists recommended that blue dyes be banned in hard candies and certain cosmetic products to reduce consumer risk. But the International Association of Color Manufacturers disagreed with the study findings, noting that the amount of dye that permeated the skin was negligible when compared with safety limits.

My family learned to avoid food dyes like the plague, reading labels like toxicologists. Everything from lip balm to ice cream became suspect, and our hypervigilance kept my daughter safe from the dangerous allergen throughout her childhood. Luckily, Benadryl comes in a dye-free formula.

I was reminded of this scary parental chapter when I wrote for the *Huffington Post* about beekeepers in France who discovered blue honey in their hives. Apparently, bees were harvesting M&M's manufacturing waste from a plant that processed the industrial runoff from a Mars candy factory. According to the BBC, the plant has now put in place a procedure to stop the problem from happening again.

Thank goodness we're not innocent bees uniquely prone to environmental pollutants. We are conscientious consumers who know real food doesn't come with a long, complex ingredient list; yet we remain vulnerable.

Here I was, fooled again, duped by new honey-coated claims. Only this time, I was polluting myself with purple dye from shampoo in hope of regaining my hair's shiny patina . . . and maintaining my sanity.

Minerals, Mollusks, and Motels

Proud that she's done her homework, my gray-haired lunch date digs into her purse and takes out a scrap of paper scribbled with notes.

"I looked it up," she says perkily. "First, apple cider vinegar to rinse off the buildup. Have you found a purple shampoo to brighten and neutralize the crazy yellow tones? Remember to double up on conditioner, and rinse a few times. Pure argon oil will moisturize any leftover dryness." She ends with, "If all else fails, eat a ton of oysters."

About ten months into my cold-turkey natural transition, I'd confessed to my friend that my two-toned dried-up hair (half dull gray from buildup and half dull dyed brown) had gotten me down . . . again. Since this was not the first time I'd dragged my well water woes and lusterless locks into our conversation, my friend headed me off at the pass with a mother lode of googled gray hair remedies. Since my own mother's voice always sings her Merry Mommy melody in my head, I can almost hear Mom chirping, with very little understanding, if any, of how the internet actually works, "You can't trust Dr. Google, you know."

After the purple shampoo debacle, I headed to the health food store, where I found an alternative. But it didn't quite get the job done, and it left my hair bone-dry. After a lifetime of having to wash my naturally normal to oily hair every day, this new thirsty hair dilemma was one I never thought I'd face. Perplexed, I thought, "How could I have let my shiny sheen devolve into a matte and muted mess?"

Oysters?

"Eating oysters will do the trick," my sunshiny friend says. "Those briny critters are rich in minerals and protein. You need nutrients for healthy hair." I give her a skeptical "Oh, really?" look. "Yes, oysters are particularly high in zinc," she pushes on. "Which aids in hair growth. The faster it grows, the sooner we can put this part of your . . . um, hair obsession to bed."

Minerals, and apparently mollusks, are swell for boosting production, but what about the environment into which the hair "seedlings" grow?

My hairdresser, Richard, tells me to get a water softener, but my husband will not get with the program.

"I've done the research, and hard water is not a health hazard. So why mess with it?" Mr. Natural argues.

Since my blood pressure and cholesterol numbers have ticked upward as I've packed on the years and a few extra pounds, I'm monitoring my salt intake. Water softeners use sodium chloride (i.e., common table salt) to soften hard water. The harder the water, the more salt needed. So a slight increase in sodium may be detected after a water softener is installed, which makes a traditional model questionable for me. The solution for my family would be to have a reverse osmosis drinking water system installed along with the water softener. Reverse osmosis removes 95 percent of "everything" in the water, including sodium.

My husband reads from the Water Research Center website about research into water hardness and cardiovascular disease: "The National Research Council (National Academy of Sciences) states that hard drinking water generally contributes a small amount toward total calcium and magnesium human dietary needs. . . . Some studies suggest a correlation between hard water and lower cardiovascular disease mortality."

Ted doesn't see the need for the water softener, doesn't want to install it, doesn't want the added cost and maintenance, and is not convinced it's a healthier alternative. He also tells me that most water softeners use polystyrene beads. The same stuff Styrofoam cups are made from.

In a decades-long marriage, two people are bound to disagree on certain issues. Ted and I email each other articles, mostly in the early morning hours when we're online reading the latest news and checking email. Often, the links to articles we send are intended to make points, and points of view, known to the other about politics, child-rearing, and health. We agree on the big stuff. It's the minutiae of everyday life that can send us on an emotional slip and slide. These emailed articles can be constructive conversation starters, or they can bring an abrupt end to a fraying discussion that's sure to go south.

Making his plastic pollution point, Ted sends me studies concluding that not only is styrene classified as carcinogenic, but polystyrene also leaches continuously from packaging and storage containers. In a study that compared Styrofoam and polystyrene cups against paper cups for safety, the Styrofoam and polystyrene cups were found to be contaminated with styrene and other aromatic compounds. While temperature played a major factor in the leaching of styrene, paper cups won out as the safer choice for hot drinks.

Without further discussion about what happens when those Styrofoam cups get dumped into the landfill, I tabled the water softener idea. Now I had to take hair matters into my own hands, which left me with two choices: go to the salon and have a stylist wash my yellowish mop with town water, or buy bottled water to shampoo with at home. A wash and blow-dry will cost me forty dollars plus tip. A supermarket gallon of bottle of water is eighty-eight cents. I opt for the cheaper route.

A few years ago, Yahoo named me one of the top ten green experts. Now I imagine myself on a tree hugger's hit list, caught red-handed, standing at the grocery store checkout line with gallons of water in my cart. Shunning eco-sensitivity, I go for it.

Back at home, I heat up the water on my gas stove, then hit the shower and stand under a forceful two-gallon stream of clear, minerally balanced bottled water. When I shampoo and condition my hair, it's like being in an Herbal Essences commercial. As the cascading mountain waterfall cradles me in an elemental downpour, a low, sexy voice in the background sings, "Take your hair to paradise and jungle-y things abound."

Hair thrives in a moisture-rich environment, and water makes up almost 25 percent of the weight of a single strand of hair. Lately, I quench that thirst by ending my shower with a cool-water rinse. According to Mireille Guiliano, author of the French Women Don't Get . . . series, rinsing hair with cool water is a French beauty secret. The lower mineral content of the soft water, along with the cool temperature, constricts hair cuticles, and the strands are left smoother and shinier.

With a viable way to obtain freshly washed hair that is glossy and manageable, I realize I can refill my water bottles at a friend's house. She's tapped into a municipal water source that runs out of her faucet like a clean mountain creek. I redeem my status as an eco-warrior.

And boy, do I enjoy an overnight visit to a swanky hotel, where I order what chef James Beard describes as "one of the supreme delights that nature has bestowed on man"—my new favorite superfood, the humble oyster. After the culinary indulgence, I step into a fragrant jungle of showering paradise.

The Landscape of Beautiful Hair

The traditional definition of beauty is "properties pleasing the eye, the ear, the intellect, the aesthetic faculty, or the moral sense." But in today's changing culture, we've narrowed our idea of beauty to an extreme focus on aesthetics—loving what we perceive as beautiful and covering up the rest, specifically with beauty products: makeup and hair dye.

On a friend's Facebook status, I recently spotted this question: "If you were a landscape, what would your landscape be?" I contemplated answering, half-jokingly, "Weathered, windswept, washed-up." Living with two-toned hair for the better part of two years opened a gate into unknown territory. I felt notably visible in a bad beauty way and invisible in an aging woman way. I didn't know what I would look like with gray hair, and my imagination ran away with unflattering stereotypes. Things were shifting, free-falling.

Taking a deep breath, I reined myself in. Every landscape tells a story of beauty: spring's botanical bounty of orange tulips sprouting up next to a pond, the return of the iridescent hummingbird to the feeder, those insightful tiny azure butterflies that lit the wooded path, and the comforting aroma of ocean air.

Weathered? Yes, the outer patina carries its age, sixty. But with age comes confidence and a grace that is fresh and surprisingly motivating. Actress Frances McDormand may have summed it up best: "Looking old . . . should be a boast about experiences accrued and insights acquired, a triumphant signal that you are someone who, beneath that white hair, has a card catalog of valuable information." And, of course, getting older is better than the alternative.

Windswept? Yes, I am carried away with meaningful work that I love and family and friends I simply adore. I can barely catch my breath while stitching it all together. I am optimistic about leaving a legacy of good.

Washed-up? Instead, I would say washed clean. Purified of the chemicals that I used to depend on for my sense of self. Rather than a stain, they have left me with a call to action.

When Rachel Carson connected chemicals used in pesticides to the death of songbirds in *Silent Spring*, she crystallized the harm we were doing to the environment and why it is essential to listen to the science: "The public must decide whether it wishes to continue on the present road, and it can do so only when in full possession of the facts. In the words of Jean Rostand, 'The obligation to endure gives us the right to know.'"

Standing at the trail's end, with the distant sound of the ferry horn across the sound, I watch Ali's plucky red pickup truck drive away from the Nature Conservancy parking lot. The wind has grown calmer as I turn back to the pond and notice a flock of my favorite cool-weather ducks, buffleheads. Bobbing their big green-and-purple heads and petite bodies while diving into the ocean for bugs and tiny fish, they coordinate flock positions. Their brilliance seems to be adapting to each and every ripple and wind shift—a tender tango. At one point during all their bobbling, these snappy-looking guys spin their heads around and switch into a carefully negotiated new dance—changing positions—ducks in front move to the middle or back, ducks in back paddle up to the front. Then they bob along, floating to a new location. As the cool spring weathers on, they are going somewhere.

~

Along with polishing my reverence for nature, working in the environmental field has opened my eyes to what is truly *not* beautiful—products filled with petrochemicals and additives that threaten to pollute our bodies. It becomes profoundly personal to realize that we have elected morally bankrupt politicians who refuse to stand up to polluters, officials who seem to care more about their careers than about the people they serve. I wish I could connect them to the beauty of nature, to make them recognize the environmental crisis as it cascades dangerously downward. But as I gaze into the mirror of the pond, I am reminded it returns only the truth of my own reflection. As Henry David Thoreau said, "It's not what you look at that matters, it's what you see." I can't make those beyond reach see what they don't want to. Yet, understanding the landscape of environmental health, how could I continue to subject my body and my planet to blatantly unregulated and under-regulated beauty products?

Our *personal nature* is elemental. Our beliefs, decisions, plans, ethics, and capacity for growth are as fluid as a glimmering deep-blue pond on a cool spring day. Growing out my chemically dyed hair reconnected me with what is clean, natural, and pure. A new, gentler landscape to love.

Chapter 10

Silver Linings

O N A SUNNY FALL DAY, Lainey and I are swinging in the hammock on my front lawn to the rhythm of a songbird chorus. Clad in light sweaters, as there is a mild chill in the air, we dangle our bare feet over the side of the hammock. Together, we flip through a copy of *Kinfolk* magazine: "The Aged Issue." When I first discovered *Kinfolk*, a hefty magazine with a highly polished vibe and a price tag to match, I thought it was pretentious, with its hipsterish outpouring of bearded guys, sullen girls, and androgynous-looking younger-than-me people. But interesting and unusual writing complements the imagery, and I was won over. Now I look forward to each themed issue. We're staring at a photo-essay titled "The Grace of Gray." After leafing through a few pages of beautiful silver-topped women, we stop to stare at the last image, of an older (than me) woman glaring back at us with a decidedly "Fuck you for thinking I'm old just because I'm gray" look.

Lainey tells me she has been noticing gray hairs and she's leaning toward coloring. I examine her head and see some silver strands around the back of her part mapping their way down her long brown curls. I don't give it much thought. Instead, I sway back and forth in the hammock, daydreaming nostalgically about all the seasons and phases of life we have ushered in. With busy lives, sometimes we're in sync, and at other times we're off in our own corners. But we're always on the same team. Our candor with each other is similar to the bond I have with my own mother. It's a treasured support inscribed in our genes.

Lainey rambles on about her grays. I am not quite paying attention until her pitch gets higher and higher and she says her gray strands don't make her feel beautiful. I ask my daughter how she knows if something is beautiful.

"I know something is beautiful if it compels me to keep looking," she answers, ever the artist that she is.

"Well, look at these women; aren't they beautiful?" I say, pointing to the page.

"They are old, and those are beautifully styled photographs, Mom." Ah, those artist eyes.

"Do you think you're beautiful?" I ask simply, not knowing if I'm opening a minefield or mindful contemplation.

She throws me one of her sideways smirks, and I fear the next sentence will start with loving sarcasm. Instead, she says, "Well, I'd feel better if I didn't have these ugly gray hairs."

Point taken. Playfully tossing back the question, she asks, "Do you think you're beautiful, Mom?"

I spurt out, "I think my favorite age is now. What choice do I have?," skirting around the answer.

Berating myself for evading the question by equating beauty to age, and thinking my parenting skills suck (granted, she's an adult), I ask myself, *Seriously, what have I learned about beauty from this gray hair adventure? Am I gray enough? Am I clean green enough? Am I mom enough?*

"We are all full of contradictions, I guess." Swinging us almost off the hammock, I shut the magazine and we head inside for wine and snacks. But the question keeps noodling away at me.

Sometimes I have to strain to remember what I looked like with a full helmet of dark hair. In my sixties now, the times when my silver head surprises me most, beyond when I pass a mirror, is when someone, usually a stranger, approaches me with a question meant as a compliment, like "Wow, is that your natural color?" Instinctively, I run my fingers through the strands and . . . voilà! . . . much to my surprise, my new friend isn't talking about the me with long, shiny black hair. It's the silver, uncolored hair that caught her eye. These encounters bring my attention back to the person I was. The one I carry with me, the one with the signature long, shiny dark natural hair.

A few days after our hammock conversation, I sent Lainey an essay written by singer Alicia Keys, with hopes of planting a seed and redeeming my parenting skills. Alicia struggles with hair, makeup, and body image.

In this viral essay, Alicia explains her feelings. "I started, more than ever, to become a chameleon." She ends the essay with a certain self-awareness and a call to action: "I don't want to cover up anymore. Not my face, not my mind, not my soul, not my thoughts, not my dreams, not my struggles, not my emotional growth. Nothing."

Middle-aged women are having a moment, and we're wearing our moment more boldly, slightly more comfortably, than before, without the veil of hair dye. Yet we're at the age when being replaced by younger, shinier versions of ourselves is the reality. And, let's face it, Alicia is in her late thirties, and her beauty naturally glows like her silky, melodic voice—fresh and clear.

Alicia's youth reminded me of something I'd just read: the luxury beauty brand Dior made the gorgeous model Cara Delevingne the face of its Capture line of wrinkle creams. Cara is twenty-five years old. What were they thinking? Twenty-five-year-olds don't need wrinkle cream. Are we older women supposed to use the cream like a diet aid and drop thirty to forty years from our appearance? The ad seems masterminded to keep women of a certain age, a group already in body image flux, insecure and covering up. And why would manufacturers change when their goal is a financial bottom line, not psychological and physical health?

It's impossible to separate aging from going gray. Natural gray hair happens. But the beauty culture seems stubbornly slow to acknowledge this intersection as companies barrage us with age-defying products, including hair dye. At every age, appearances matter. Staying attractive is important, and how we get and stay there is equally important. Life is a long game. Wanting to be in it for the duration, journalist Linda Ellerbee said, "My friend, Gloria Steinem, taught me how to age. Basically she'd sum it up as: 'Don't stop.'"

Exactly; what choice do we have? To which I should have added, when answering my beautiful daughter, "What choice do we have, once we make decisions about our beauty with health as the compass?" In the spirit of self-discovery, I'm giving myself a do-over. The real answer to Lainey is, "I look as realistically beautiful as a *healthy* woman can. This is me. I know more about who I am now. I like it."

At the beginning of my going natural journey, I often said, "If gray hair makes me feel old, I can always go back to coloring." I've finally stopped backpedaling. I realize now that I won't color again because I can never unlearn what I've discovered about the matrix of hair color and health.

Nearing the end of writing this book, I learned about a woman who worked at a law firm that specialized in asbestos litigation. She was working with a lab on a toxics case when she decided to have her six-year-old daughter's "play makeup" tested. Much to her horror, she discovered it contained asbestos.

"You assume that when you're purchasing it, it is safe. I remember literally sinking to the ground just being like, 'Oh my gosh!'" Kristina Warner said in a radio interview. She had bought the makeup for her daughter at Claire's, the popular mall jewelry and accessory store for girls and tweens.

Claire's pulled the items from shelves after learning of the findings. But how many other six-year-olds were exposed?

Here's what I know: Human nature tells us to protect—to protect ourselves and our children. In this vast tapestry of life, it's a basic human need. We can't bank on profit-seeking companies to do it. We can't wait around for our government officials to play nice and work together. History tells us they rarely do. Yet it's not fair to dump responsibility in the laps of consumers. Not just because it's unethical but because we all can't be litigator moms working in science labs, or investigators writing reports and studies, or full-time environmental health advocates lobbying on Capitol Hill.

I thought a lot about whether or not to include politics in this hair story. We all know politics is polarizing. It's also a moving target. But here's the thing: in this particular space and time, politics and protecting human health are inseparable. Those without a degree in toxicology have almost no way to determine what is safe and what is not in hair dye. To leave it up to consumers to find answers to complicated health issues is absolutely unjust. And women often bear the brunt of this injustice. We are disproportionately, far more than men, victims of body pollution and therefore of diseases linked to our environment.

While we can't decipher every ingredient in every product, there are things we can do to move closer to a nontoxic world. First, we can advocate for sound science that clarifies the risk of chemical dyes. Additional research is needed to create strategies to avoid toxic chemical exposure at home and in salons. This will lead to a better understanding of hairdressers' health and women's health in general, including the effects of collective exposures to multiple chemicals. We deserve this research for ourselves and for future generations, who can't advocate for it themselves.

Scientific research is important, but it is only part of the train ride that is chemical reform. Knowledge is a passenger transported from place to place, one report, one study, and one person at a time. As citizens, we are the locomotion conductors. We're driving change by carrying knowledge into the world. It can be a dizzying and dispiriting job in this day and age of misinformation. We are already hitting brick walls or coming to a screeching halt even before the train leaves the station.

What else can consumers of hair and beauty products do? If we don't buy shampoos, dyes, and cosmetics with questionable ingredients, manufacturers will clamor to create healthier ones to meet customer demand. We can boycott and vote with our pocketbooks.

Along with becoming aware of body pollution, we must learn to spot political pollution, to recognize when special interests are poisoning our regulatory policy for their financial benefit. Once unleashed, this corruption becomes a vicious and unruly beast. Defeating it may become the fight of our times as we advocate for stronger health protections.

We still have a lot of work to do. Ensuring that our lawmakers act in our best interest is ultimately the goal. Yes, this can be exhausting; at times, it feels like progress is nearly impossible. But legislators need to hear from their constituents over and over again (resist, resist). They need to know that we'll vote them out of office if they are unwilling to pass strong laws to ensure that our products are safer and our health is protected. Regardless of which party is in office, all lawmakers need to hear from citizens often. At this pivotal time, each and every one of us can play a role in democracy if we exercise our right to speak out.

So, here I am, living the consequence of my ecological awareness, reckoning with the interplay between health and beauty. One of the steepest learning curves for me was realizing that being anxious about toxic chemicals doesn't necessarily mean you won't use them. Facts can't always compete with ingrained beauty ideals. It may not seem logical, but that's human nature. I am guilty of it too, as I have recently reverted to using that damn purple shampoo if I wash my hair with mineral-laden water. We keep using the stuff because we rationalize that changing our appearance will change our life. We fear that giving up that special product, whatever it is, will render us unattractive or irrelevant. Of course, this malcontent keeps the beauty industry in the business of targeting women with campaigns that perpetuate hair and beauty myths.

With a few setbacks, I have learned to quiet those fears and make going gray my new "hair is life" story. I've come to realize it's my attitude that has changed as much as my hair. Because beauty, unlike nature, is something we create. It ebbs and flows with how intensely we focus on it . . . how awakened our minds are . . . how open our hearts are to change. I was finally able to get far enough away from the hair dyeing culture that I stopped worrying about returning to it.

I used to see hair dye as the be-all and end-all for camouflaging age, but now I see it as a dead end. Our ideas about turning back the hands of time necessarily shift as we learn more and more about what keeps us healthy. Attitudes and minds change, and in that sense, for me, going gray created a challenge to be stunningly honest, to transform not only my hair but also my sense of self, to accept the person within.

As I write these last pages, sitting in solitude, witnessing nature's changes from my screened-in porch, I stop often to gaze at the peeling birch trees. Over the past thirty years, I have watched them transition from young saplings to tall, majestic beauties. There is nothing ordinary about them—sleek pillars of grace, full of longing as they stretch toward the sky. Stark against nature's greens and blues, their chalky silver bark peels off in long, deliberate strips. Over the years, their outer layers have gone from silky smooth to lightly ridged as the trees' hidden layers mature. If the birches could talk, would they say their perspective has changed? Would they say they're following the oaks and maples crowd? Would they say they've given up the gravitational force to soar? Without worry of being sprayed, maimed, or cut down in their prime, the trees stand tall, their leaves fluttering in the breeze, whispering freedom. For the birches, natural beauty is not a contradiction of age or time. They keep climbing. Descent is not an option. Continuing to sprout leaves year after year, their dazzling bark magnifies a new story, one I've come to know as my own crowning glory.

Acknowledgments

First, I would like to thank Dominique Browning. For years, I have been inspired by her compassionate writing. But it wasn't until she entrusted me as editor of her new project, Moms Clean Air Force, that my writing wings fully took flight. Many people encouraged me to go gray, but Dominique believed I could write a book that could not only share the ins and outs of this particular beauty choice but also add value to one of the most important environmental stories of our time: the health effects of potentially harmful chemicals in personal care products, such as hair dye.

Thank you to my editor, Emily Turner. She was insightful, wise, and patient. Emily does what every good editor does: she makes a book better. Heartfelt thanks to the Island Press team for getting *True Roots* out into the world.

Deep appreciation to my literary agent, Diana Finch. From the start, Diana firmly grasped the concept of the book. Her enthusiasm helped me stay on course in the belief that women of all ages want and need this book.

A special thank-you to my dearest and oldest friends, Cathy and Ralph Segalowitz, who read through the early drafts and provided quick and expert feedback. Thanks are also due to family, friends, and colleagues at Moms Clean Air Force and the Environmental Defense Fund who supported me throughout the journey and tolerated the bemoaning when my hair obsession threatened to throw me off course.

My dream of writing a book by the ocean began at the Martha's Vineyard Writer's Residency at the Noepe Center for Literary Arts. Thank you, Justen Ahren, Jack Sonni, Sean Murphy, and the residents who shared their writing life wisdom with me. Many thanks to Laurie David, who continued my dream by introducing me to Iya Labunka. I am grateful to Iya for opening her home to me. My writing season by the sea (nor'easters and all) brought the calming element of the natural world into the difficult work of understanding how business and politics conspire in the most unnatural ways.

Above all, thank you to my husband, Ted, who is my sounding board and staunch going gray supporter; my daughter, Lainey, and her husband, Ben Scott; my son, Ben, and his wife, Ella Maslin, for all their unwavering love; and my mother, Joan Citron, whose indomitable spirit has always been a pillar of strength for me. A girl's hair may be her life, but family is everything.

Bibliography

Abelman, Devon. "Gray Hair Is Set to Be 2018's Most Popular Hair-Color Trend." *Allure*, January 5, 2018. https://www.allure.com/story/tracing-the-gray-hair-trend.

Ahmad, Maqbool, and Ahmad S. Bajahlan. "Leaching of Styrene and Other Aromatic Compounds in Drinking Water from PS Bottles." *Journal of Environmental Science* 19, no. 4 (2007): 421–426. doi:10.1016/S1001–0742(07)60070–9.

Akst, Jef. "The Elixir Tragedy, 1937." *The Scientist*, June 1, 2013. https://www.the-scientist.com/foundations/the-elixir-tragedy-1937-39231.

Alexander, Ty. "Why Does Grey Hair Turn Yellow + 4 Home Remedies to Remove That Yellow Tinge from Your Hair." *Gorgeous in Grey* (blog). Accessed December 5, 2018. https://gorgeousingrey.com/home-remedies-remove-yellow-gray-hair/.

Amarelo, Monica. "Environmental Groups Sue FDA to Take Formaldehyde Out of Salons." Environmental Working Group, August 7, 2017. https://www.ewg.org/release/environmental-groups-sue-fda-take-formaldehyde-out-salons#.WiFu4baZOu7.

American Cancer Society. "Hair Dyes." Last modified May 27, 2014. https://www.cancer.org/cancer/cancer-causes/hair-dyes.html.

American Chemical Society. "Blackcurrant Dye Could Make Hair Coloring Safer, More Sustainable." *ScienceDaily*, May 29, 2018. Accessed November 9, 2018. https://www.sciencedaily.com/releases/2018/05/180529103538.htm.

American Psychological Association. "Understanding How People Change Is First Step in Changing Unhealthy Behavior." December 3, 2003. http://www.apa.org/research/action/understand.aspx.

Appelbaum, Binyamin, and Jim Tankersley. "The Trump Effect: Business, Anticipating Less Regulation, Loosens Purse Strings." *New York Times*, January 1, 2018. https://www.nytimes.com/2018/01/01/us/politics/trump-businesses-regulation-economic-growth.html.

Archie Comics. "Veronica Lodge." Accessed October 31, 2018. http://archiecomics.com/characters/veronica-lodge/.

Arnett, Kristen. "I Tried Hairprint—the Safest Way to Cover Grey Hair—Here's What Happened." Accessed July 20, 2018. https://kristenarnett .com/hairprint-review-safest-way-cover-grey-hair/.

Baby Tooshy. "Disposable Diapers Add Millions of Tons of Waste to Landfills Each Year, According to EPA Report." Cision PR Newswire, December 31, 2016. https://www.prnewswire.com/news-releases/ disposable-diapers-add-millions-of-tons-of-waste-to-landfills-each -year-according-to-epa-report-300384344.html.

Bairstow, Jonny. "Dye-ing to Have Sustainably Coloured Hair? Blackcurrant Waste Could Help." Energy Live News, May 31, 2018. https:// www.energylivenews.com/2018/05/31/dye-ing-to-have-sustainably -coloured-hair-blackcurrant-waste-could-help/.

Bazilian, Emma. "Why Older Women Are the New It-Girls of Fashion." *Adweek*, April 6, 2015. http://www.adweek.com/brand-marketing/why -older-women-are-new-it-girls-fashion-163871/.

Beard, Mary. "Vanity Isn't to Blame for Our Addiction to Hair Dye, Insists TV Historian Mary Beard, Who Says Women Are Victims of a Great Grey Hair Conspiracy." *Daily Mail*, February 24, 2016. http:// www.dailymail.co.uk/femail/article-3463108/Vanity-ISN-T-blame -addiction-hair-dye-insists-TV-historian-Mary-Beard-says-women -victims-great-grey-hair-conspiracy.html.

Becker, Katie. "10 American Beauty Ingredients That Are Banned in Other Countries." *Cosmopolitan*, November 8, 2016. https://www.cosmopolitan .com/style-beauty/beauty/g7597249/banned-cosmetic-ingredients/.

Benedict, Elizabeth, ed. *Me, My Hair, and I: Twenty-Seven Women Untangle an Obsession.* Chapel Hill: Algonquin Books, 2015.

Berlinger, Max. "For Millennial Men, Gray Hair Is Welcome." *New York Times*, February 3, 2016. https://www.nytimes.com/2016/02/04/ fashion/for-millennial-men-gray-hair-is-welcome.html.

Bienkowski, Brian. "It's Time to Rethink Chemical Exposures—'Safe' Levels Are Doing Damage: Study." Environmental Health News, December 20, 2017. http://www.ehn.org/chemical-exposures-are-small -doses-harm-2518446452.html.

Bienkowski, Brian. "Only Half of Drugs Removed by Sewage Treatment." *Scientific American*, November 22, 2013. https://www.scientificamerican .com/article/only-half-of-drugs-removed-by-sewage-treatment/.

Black Girl Health (blog). "Drinking Water for Healthy Hair." Accessed December 5, 2018. http://blackgirlhealth.com/portfolio/drinking-water -for-healthy-hair/.

Blackburn, Jo Glanville. "As Health Concerns Mount over Conventional Hair Dyes, Is It Time to Revisit a Seventies Classic? Henna's Back—but Not as You Knew It." *Daily Mail*, April 29, 2018. http://www.dailymail.co.uk/femail/article-5671719/Hennas-not-knew-it.html.

Bloom, Ester. "Here's How Much the Average American Spends on Health Care." *CNBC Make It*, June 23, 2017. https://www.cnbc.com/2017/06/23/heres-how-much-the-average-american-spends-on-health-care.html.

Bloomberg. "David Lewis, Green Chemicals PLC: Profile & Biography." Accessed November 1, 2018. https://www.bloomberg.com/profiles/people/16547369-david-m-lewis.

Brain, Marshall. "How Hair Coloring Works." How Stuff Works. Last modified April 1, 2000. https://science.howstuffworks.com/innovation/everyday-innovations/hair-coloring3.htm.

Brazilian Blowout. "Professional Hair Smoothing Treatment & Keratin Hair Smoothing Treatment." Accessed October 29, 2018. https://www.brazilianblowout.com.

Brody, Jane E. "For Cosmetics, Let the Buyer Beware." *New York Times*, August 7, 2017. https://www.nytimes.com/2017/08/07/well/for-cosmetics-let-the-buyer-beware.html?_r=0.

Browning, Dominique. "The Wind Map: We Share the Air." Moms Clean Air Force, May 1, 2012. https://www.momscleanairforce.org/the-wind-map-we-share-the-air/.

Cama, Timothy. "Trump Admin Appeals Ruling Ordering EPA to Ban Pesticide." *The Hill*, September 24, 2018. https://thehill.com/policy/energy-environment/408173-trump-admin-appeals-ruling-ordering-epa-to-ban-pesticide.

Campaign for Safe Cosmetics. "Nitrosamines." Accessed November 1, 2018. http://www.safecosmetics.org/get-the-facts/chemicals-of-concern/nitrosamines/.

Carrington, Damian. "Plastic Fibres Found in Tap Water around the World, Study Reveals." *Guardian*, September 5, 2017. https://www.theguardian.com/environment/2017/sep/06/plastic-fibres-found-tap-water-around-world-study-reveals.

Carroll, Linda. "Chemicals in Vinyl Flooring and Wallpaper Raise Worries." Children's Health on NBC News.com, November 22, 2010. http://www.nbcnews.com/id/39728598/ns/health-childrens_health/t/chemicals-vinyl-flooring-wallpaper-raise-worries/#.WqBlCWaZMdU.

CBS WCCO Minneapolis. "Uptown Hair Salon Goes Green, Diverts 95 Percent of Waste." CBS Minnesota, January 16, 2015. http://minnesota .cbslocal.com/2015/01/16/uptown-hair-salon-goes-green-diverts-95 -percent-of-waste/.

The Center for Responsive Politics. "Trump 2017 Inauguration Donors." OpenSecrets.org. Accessed October 31, 2018. https://www.opensecrets .org/trump/inauguration-donors.

Centers for Disease Control and Prevention. "Breast Cancer Rates among Black Women and White Women." Accessed October 29, 2018. https:// www.cdc.gov/cancer/dcpc/research/articles/breast_cancer_rates _women.htm.

Chemical Inspection and Regulation Services. "EU Cosmetic Regulations and Registration." Accessed October 31, 2018. http://www.cirs-reach .com/Cosmetics_Registration/eu_cosmetics_directive_cosmetics _registration.html.

Cherry, Kendra. "The 6 Stages of Behavior Change: The Transtheoretical or Stages of Change Model." VeryWell Mind, updated November 18, 2018. https://www.verywellmind.com/the-stages-of-change-2794868.

Choma, Russ, and Robbie Feinberg. "Chemical, Pharmaceutical Industries See Huge Lobbying Increases." OpenSecrets.org, May 2, 2014. http://www.opensecrets.org/news/2014/05/chemical-pharmaceutical -industries-see-huge-lobbying-increases/.

Chong, H. P., K. Reena, Y. N. Khuen, R. Y. Koh, H. N. Chew, and S. M. Chye. "Para-Phenylenediamine Containing Hair Dye: An Overview of Mutagenicity, Carcinogenicity, and Toxicity." *Journal of Environmental & Analytical Toxicology* 6 (2016): 403. doi:10.4172/2161-0525.1000403.

Chow, Lorraine. "EPA Chief Met with Dow Chemical CEO Before Deciding Not to Ban Toxic Pesticide." EcoWatch, June 28, 2017. https:// www.ecowatch.com/pruitt-dow-chlorpyrifos-2449662128.html.

Citron-Fink, Ronnie. "Don't Let Politicians Pollute the EPA." Moms Clean Air Force, February 13, 2017. http://www.momscleanairforce .org/life-before-epa/.

Citron-Fink, Ronnie. "Only Fools Dye Their Young." *Huffington Post*, October 8, 2012. https://www.huffingtonpost.com/entry/only-fools-dye -their-youn_b_1946584.html.

Coggon, Matthew M., Brian C. McDonald, Alexander Vlasenko, Patrick R. Veres, François Bernard, Abigail R. Koss, Bin Yuan, et al. "Diurnal Variability and Emission Pattern of Decamethylcyclopentasiloxane

(D5) from the Application of Personal Care Products in Two North American Cities." *Environmental Science & Technology* 52, no. 10 (2018): 5610–5618. doi:10.1021/acs.est.8b00506.

Coggon, Matthew, and Karin Vergoth. "Personal Plumes: New Study: Daily Emissions from Personal Care Products Comparable to Car Emissions, Contribute to Air Pollution in Boulder." Cooperative Institute for Research in Environmental Sciences, April 30, 2018. https://cires.colorado.edu/news/personal-plumes.

Coleman, Claire. "Why Are Today's Women Going Grey at 25?" *Daily Mail*, April 4, 2011. http://www.dailymail.co.uk/femail/article-1368239/Why-todays-women-going-grey-25.html.

Columbia Research. "Safely Handling Dyes." Columbia University in the City of New York. https://research.columbia.edu/content/safely-handling-dyes.

Conway, Beth. "6 Myths about Toxic Chemicals to Quash Today!" Women's Voices for the Earth, March 30, 2018. https://www.womensvoices.org/2018/03/30/6-myths-about-toxic-chemicals-to-quash-today/.

Cosmetic Ingredient Review. "Currently Under Review." Accessed October 31, 2018. https://www.cir-safety.org.

Cowles, Charlotte. "First Looks: 'Schiaparelli and Prada: Impossible Conversations' at the Met Costume Institute." The Cut, May 7, 2012. https://www.thecut.com/2012/05/first-looks-the-mets-schiaparelli-and-prada-costume-instute.html.

Creswell, Julie. "Is It 'Natural'? Consumers, and Lawyers, Want to Know." *New York Times*, February 16, 2018. https://www.nytimes.com/2018/02/16/business/natural-food-products.html.

Dahl, Lindsay. "Regulation Is Not a Dirty Word." LindsayDahl.com, January 31, 2017. http://www.lindsaydahl.com/regulations-that-protect-our-health/.

Daniluk, Julie. "Oysters for Healthy Hair + Other Beautifying Foods." MindBodyGreen, October 5, 2018. https://www.mindbodygreen.com/0-16435/oysters-for-healthy-hair-other-beautifying-foods.html.

Daswani, Kavita. "More Men Coloring Their Hair." *Los Angeles Times*, January 29, 2012. http://articles.latimes.com/2012/jan/29/image/la-ig-mens-hair-color-20120129.

Dawson, Katherine. "Riccardi on Quest to Find Best Toxic-Free Hair Products." *Greenwich (CT) Sentinel*, July 30, 2018. https://www.greenwich

sentinel.com/2018/07/30/riccardi-on-quest-to-find-best-toxic-free-hair
-products/.

Denchak, Melissa. "All about Alar." National Resources Defense Council. Last modified March 14, 2016. https://www.nrdc.org/stories/all-about
-alar.

Diaz, Jesus. "Graphene Hair Dye Is Coming, and It Looks Incredible." *Fast Company*, March 16, 2018. https://www.fastcodesign.com/90164386/
graphene-hair-dye-is-coming-and-it-looks-incredible.

DK. *Natural Beauty*. New York: DK Publishing, 2015.

Dove. "About Dove." Accessed October 31, 2018. http://www.dove.us/
Social-Mission/Self-Esteem-Statistics.aspx.

"Do You Know What's in Your Cosmetics?" Editorial. *New York Times*, February 9, 2019. https://www.nytimes.com/2019/02/09/opinion/
cosmetics-safety-makeup.html.

EarthTalk. "Bad Hair Day: Are Aerosols Still Bad for the Ozone Layer?" *Scientific American* online. Accessed October 31, 2018. https://www
.scientificamerican.com/article/are-aerosols-still-bad/.

EarthTalk. "Eco-Dos: Green Beauty Salons and Hair Products Are a Growing Business." *Scientific American* online. Accessed November 9, 2018. https://www.scientificamerican.com/article/earth-talks-eco
-dos/.

Ellen MacArthur Foundation. "A New Textiles Economy: Redesigning Fashion's Future." Accessed December 4, 2018. https://www.ellenmacarthur
foundation.org/assets/downloads/publications/A-New-Textiles-Economy
_Full-Report_Updated_1-12-17.pdf.

Environmental Defense Fund. "Victory for Advocates on Lead Acetate: FDA Agrees to Ban Toxic Lead Compound from Hair Dyes." October 30, 2018. https://www.edf.org/media/victory-advocates-lead-acetate
-fda-agrees-ban-toxic-lead-compound-hair-dyes.

Environmental Working Group (EWG). "EWG: Lipstick Chemicals Declared Toxic by Canadian Gov't." Last modified January 31, 2009. https://www.ewg.org/news/news-releases/2009/02/02/ewg-lipstick
-chemicals-declared-toxic-canadian-gov't#.Wux1Ay-ZP-Y.

Environmental Working Group (EWG). "Personal Care Products Safety Act (S. 1113) Summary." Accessed December 5, 2018. https://cdn3
.ewg.org/sites/default/files/u352/Personal%20Care%20Products%20
Safety%20Act%20%28S.1113%29%20Summary%20.pdf?_ga=2.17079737
.1858962705.1543446638-1309551093.1425498458.

Environmental Working Group (EWG). "Personal Care Products Safety Act Would Improve Cosmetics Safety." Accessed December 5, 2018. https://www.ewg.org/Personal-Care-Products-Safety-Act-Would-Improve-Cosmetics-Safety.

Environmental Working Group (EWG) Skin Deep Cosmetic Database. "Chemicals in Cosmetics Linked to Lung Damage in Children." Accessed February 7, 2019. https://www.ewg.org/release/study-chemicals-cosmetics-linked-lung-damage-children.

Environmental Working Group (EWG) Skin Deep Cosmetic Database. "Clairol Hair Dye." Accessed December 5, 2018. https://www.ewg.org/skindeep/search.php?query=clairol+hair+dye.

Environmental Working Group (EWG) Skin Deep Cosmetic Database. "D&C Violet 2." Accessed November 9, 2018. http://www.ewg.org/skindeep/ingredient/701828/D%26C_VIOLET_2/.

Environmental Working Group (EWG) Skin Deep Cosmetic Database. "Just For Men." Accessed October 29, 2018. https://www.ewg.org/skindeep/brand/Just_For_Men/#.Wn9hkSOZP-Z.

Ephron, Nora. *I Feel Bad about My Neck*. New York: Knopf, 2006.

Erlich, Jessica Prince. "Secrets to Gorgeous Hair Color." *Harper's Bazaar*, November 14, 2016. https://www.harpersbazaar.com/beauty/hair/advice/a5586/home-hair-color-1110/.

Esposito, Lisa. "Are You Allergic to Hair Dye?" *U.S. News & World Report*, May 15, 2017. https://health.usnews.com/health-care/patient-advice/articles/2017-05-15/are-you-allergic-to-hair-dye.

Fennessy, Lisa. "Organic Hair Dyes: The Good. The Bad. The Ugly." *This Organic Girl* (blog), January 22, 2016. https://thisorganicgirl.com/organic-hair-dye-the-good-the-bad-and-the-ugly/.

Fisher, Michele. "Use of Dark Hair Dye and Relaxers Associated with Increased Breast Cancer Risk." Rutgers Cancer Institute of New Jersey: Rutgers Health, June 14, 2017. http://cinj.org/use-dark-hair-dye-and-relaxers-associated-increased-breast-cancer-risk.

Flint, Nourbese N., and Teniope Adewumi. "Natural Evolutions: One Hair Story." Black Women for Wellness, 2016. http://www.bwwla.org/newsite2017/wp-content/uploads/2017/03/One-Hair-Story-Final-small-file-size-3142016.pdf.

Frazier, Ian. "Can Sylvia Earle Save the Oceans?" *Outside*, November 12, 2015. https://www.outsideonline.com/2030946/marine-biologist-sylvia-earle-profile.

Fredericks, Jonae. "How to Remove Yellow from White Hair Naturally." Livestrong.com, July 18, 2017. http://www.livestrong.com/article/226984-how-to-remove-the-yellow-from-white-hair-naturally/.

Gago-Dominguez, Manuela, J. Esteban Castelao, Jian-Min Yuan, Mimi C. Yu, and Ronald K. Ross. "Use of Permanent Hair Dyes and Bladder-Cancer Risk." *International Journal of Cancer* 91, no. 4 (December 2000): 575–579. https://www.ncbi.nlm.nih.gov/pubmed/11251984.

Ghilarducci, Teresa. "The Unique Disadvantage Older Women Face in the Workforce." *PBS NewsHour*, March 25, 2016. https://www.pbs.org/newshour/economy/the-unique-disadvantage-older-women-face-in-the-workforce.

Gladwell, Malcolm. "True Colors." *New Yorker*, March 22, 1999. http://www.newyorker.com/magazine/1999/03/22/true-colors.

"Global Hair Care Markets." November 2018. ReportLinker.com. https://www.reportlinker.com/do121808245/Global-Hair-Care-Markets.html.

Godin, Seth. "The Pleasure/Happiness Gap." *Seth's Blog*, October 1, 2017. http://sethgodin.typepad.com/seths_blog/2017/10/the-pleasure-happiness-gap.html.

Goop. "Eight Rules for Safer Hair Color." Accessed October 29, 2018. https://goop.com/beauty/hair/eight-rules-for-safer-hair-color/.

Green Circle Salons. *Make Beauty Beautiful: The Whole Story*. Green Circles Salon video, 1:09 min. Accessed November 9, 2018. https://greencirclesalons.com.

Grossman, Elizabeth. *Chasing Molecules: Poisonous Products, Human Health, and the Promise of Green Chemistry*. Washington, DC: Island Press, 2012.

Guenard, Rebecca. "Down the Drain." Mosaic, December 16, 2014. https://mosaicscience.com/story/down-drain/.

Guenard, Rebecca. "Hair Dye: A History." *The Atlantic*, January 2, 2015. https://www.theatlantic.com/health/archive/2015/01/hair-dye-a-history/383934/.

Guiliano, Mireille. "5 Top Beauty Secrets of French Women." French Women Don't Get Fat. Accessed November 1, 2018. http://frenchwomendontgetfat.com/5-top-beauty-secrets-of-french-women/.

Hacinli, Cynthia. "Safer Hair Color?" *Washingtonian*, September 1, 2008. https://www.washingtonian.com/2008/09/01/safer-hair-color/.

Hairfinder. "History of Haircolor." Accessed October 31, 2018. https://www.hairfinder.com/info/haircolor-history.htm.

Hairprint. "We Care." Accessed November 9, 2018. https://www.myhairprint.com/pages/we-care.

Hawken, Paul. "Drawdown." Accessed November 9, 2018. http://www
.paulhawken.com.

Hawken, Paul, ed. *Drawdown: The Most Comprehensive Plan Ever Proposed to Reverse Global Warming*. New York: Penguin Books, 2017.

Haynes, Chelsea. "True Cost of Beauty: Survey Reveals Where Americans Spend Most." Groupon Merchant, August 3, 2017. https://www.groupon .com/merchant/blog/true-cost-beauty-americans-spend-most-survey.

HealthDay. "Lead in Hair Dyes Must Go: FDA." October 30, 2018. https://consumer.healthday.com/women-s-health-information-34/ cosmetic-news-158/lead-in-hair-dyes-must-go-fda-739111.html.

Hefford, Bob. "Colour and Controversy." *Chemistry World*, October 3, 2012. https://www.chemistryworld.com/opinion/colour-and-controversy/ 5479.article.

Holland, Kimberly. "Get the Lead Out . . . of Men's Hair Dye." Health-line, April 12, 2017. https://www.healthline.com/health-news/get-the -lead-out-of-mens-hair-dye#2.

Hopp, Deven. "From 1500 BC to 2015 AD: The Extraordinary History of Hair Color." *Byrdie*, December 9, 2015. https://www.byrdie.com/hair -color-history/slide2.

Hou, Kathleen. "There Could Finally Be a Bill Regulating the Safety of Beauty Products." The Cut, May 11, 2017. https://www.thecut.com/2017/ 05/the-personal-care-products-safety-act-will-regulate-beauty.html.

Howard, Brian Clark. "Q&A: Sylvia Earle's Personal Journey and Why the Ocean Is Vital to Life." *National Geographic*, August 16, 2014. https://news .nationalgeographic.com/news/2014/08/140815-sylvia-earle-mission-blue -documentary-film-ocean-environment/.

Hubbard, Lauren. "These 8 Models Will Give You Gray Hair Goals." *Town and Country*, October 21, 2016. https://www.townandcountrymag .com/style/beauty-products/g2959/gray-hair-models/.

Huen, Karen, Asa Bradman, Kim Harley, Paul Yousefi, Dana Boyd Barr, Brenda Eskenazi, and Nina Holland. "Organophosphate Pesticide Levels in Blood and Urine of Women and Newborns Living in an Agricultural Community." *Environmental Research* 117 (August 2012): 8–16. doi:10.1016/j.envres.2012.05.005.

Intergovernmental Panel on Climate Change. "Press Release: Human Influence on Climate Clear, IPCC Report Says." September 27, 2013. http://www.climatechange2013.org/images/uploads/WGI-AR5_SPM PressRelease.pdf.

International Association of Color Manufacturers. "International Association of Color Manufacturers (IACM)." Accessed November 9, 2018. https://www.iacmcolor.org.

Island Grown Initiative. "Our Staff." Accessed November 9, 2018. https://www.igimv.org/about/staff-board.

Jaouad, Suleika. "Hair, Interrupted." *Vogue*, May 28, 2015. https://www.vogue.com/article/hair-loss-cancer-stories-chemotherapy-suleika-jaouad.

J Crew. "Beauty Notes: Model Pia Gronning, 65." *JCrew Blog*. Accessed November 28, 2018. https://hello.jcrew.com/2015-02-mar/beauty-notes-meet-pia-gronning.

Jensen, Michael C. "Tobacco: A Potent Lobby." *New York Times*, February 19, 1978. https://www.nytimes.com/1978/02/19/archives/tobacco-a-potent-lobby-lobby.html.

Journal of Personality and Social Psychology. [Report on women's self-esteem, 2015.] Accessed December 5, 2018. http://www.apa.org/pubs/journals/psp/index.aspx.

Kaplan, Sheila. "In Search of the Perfect Hair Dye." *New York Times*, March 16, 2018. https://www.nytimes.com/2018/03/16/science/hair-dye-graphene.html.

Keys, Alicia. "Alicia Keys: Time to Uncover." *Lenny* (blog), May 31, 2016. http://www.lennyletter.com/style/a410/alicia-keys-time-to-uncover/.

Kreamer, Anne. *Going Gray: How to Embrace Your Authentic Self with Grace and Style*. New York: Little, Brown, 2009.

Kristen. "Earth Day Primpin': Eco-Activist Paul Hawken Talks Beauty Biz and His Own Regime." *Holistic Vanity* (blog), May 1, 2017. http://www.holisticvanity.ca/earth-day-primpin-eco-activist-paul-hawken-talk/.

Kristof, Nicholas. "Contaminating Our Bodies with Everyday Products." *New York Times*, November 28, 2015. https://www.nytimes.com/2015/11/29/opinion/sunday/contaminating-our-bodies-with-everyday-products.html?smid=fb-share.

Kristof, Nicholas. "Trump's Legacy: Damaged Brains." *New York Times*, October 28, 2017. https://www.nytimes.com/interactive/2017/10/28/opinion/sunday/chlorpyrifos-dow-environmental-protection-agency.html?emc=edit_th_20171029&nl=todaysheadlines&nlid=55963803.

La Ferla, Ruth. "Not Selling Gray Hair Short." *New York Times*, January 25, 2013. http://www.nytimes.com/2013/01/25/booming/not-selling-gray-hair-short.html.

Lanphear, Bruce P. "Low-Level Toxicity of Chemicals: No Acceptable Levels?" *PLoS Biology* 15, no. 12 (December 19, 2017): e2003066. doi:10.1371/journal.pbio.2003066.

Lee, Michelle. "*Allure* Magazine Will No Longer Use the Term 'Anti-Aging.'" *Allure*, August 14, 2017. https://www.allure.com/story/allure-magazine-phasing-out-the-word-anti-aging.

Leone, Jared. "Claire's Pulls Makeup with Asbestos Contamination." *Atlanta Journal-Constitution*, January 5, 2018. https://www.ajc.com/news/national/claire-pulls-makeup-with-asbestos-contamination/lPiLKAoQ8Y4OvyxVP9kHuL/.

Lipton, Eric. "The Chemical Industry Scores a Big Win at the E.P.A." *New York Times*, June 7, 2018. https://www.nytimes.com/2018/06/07/us/politics/epa-toxic-chemicals.html.

Lipton, Eric. "F.D.A. Has 6 Inspectors for 3 Million Shipments of Cosmetics." *New York Times*, August 2, 2017. https://www.nytimes.com/2017/08/02/us/politics/fda-has-6-inspectors-for-3-million-shipments-of-cosmetics.html?emc=edit_th_20170803&nl=todaysheadlines&nlid=55963803&_r=0.

Llanos, Adana A. M., Anna Rabkin, Elisa V. Bandera, Gary Zirpoli, Brian D. Gonzalez, Cathleen Y. Xing, Bo Qin, et al. "Hair Product Use and Breast Cancer Risk among African American and White Women." *Carcinogenesis* 38, no. 9 (September 1, 2017): 883–892. doi:10.1093/carcin/bgx060.

Local Hazardous Waste Management Program in King County, Washington. "Household Hazardous Products List." https://www.hazwastehelp.org/HHW/list.aspx.

Luo, Chong, Lingye Zhou, Kevin Chiou, and Jiaxing Huang. "Multifunctional Graphene Hair Dye." *Chem* 4, no. 4 (April 12, 2018): 784–794. doi:10.1016/j.chempr.2018.02.021.

Mandal, Ananya. "Fungal Toxins from Wallpaper Source of Illness Says New Research." News-Medical.net, June 25, 2017. https://www.news-medical.net/news/20170625/Fungal-toxins-from-wallpaper-source-of-illness-says-new-research.aspx.

Mandell, Janna. "L'Oréal Goes Vegan; Announces New Plant-Based Hair Color Line." *Forbes*, December 13, 2017. https://www.forbes.com/sites/jannamandell/2017/12/13/loreal-goes-vegan-announces-new-plant-based-hair-color-line/#23ee3de94c5b.

Marshall, Sarah. "When, and Why, Did Women Start Dyeing Their Gray Hair?" *Elle*, September 18, 2015. https://www.elle.com/beauty/

hair/news/a30556/when-and-why-did-women-start-dyeing-their-gray
-hair/.

Maynard, Andrew. "Eager to Dye Your Hair with 'Nontoxic' Graphene Nanoparticles? Not So Fast!" *The Conversation*, March 20, 2018. http:// theconversation.com/eager-to-dye-your-hair-with-nontoxic-graphene -nanoparticles-not-so-fast-93523.

McDonough, William, and Michael Braungart. *Cradle to Cradle: Remaking the Way We Make Things.* New York: North Point Press, 2002.

McLintock, Kaitlyn. "Here's Everything You Need to Know about Using Organic Hair Dye." *Byrdie*, December 15, 2017. https://www.byrdie .com/organic-hair-dye.

Mendes, Elizabeth. "More than 4 in 10 Cancers and Cancer Deaths Linked to Modifiable Risk Factors." American Cancer Society, November 21, 2017. https://www.cancer.org/latest-news/more-than-4-in-10-cancers -and-cancer-deaths-linked-to-modifiable-risk-factors.html.

Mills, Danielle. "Salon Waste Management." *Eco Hair and Beauty*, June 25, 2015. http://ecohairandbeauty.com/salon-waste-management/.

Mintel Press Team. "Mintel Announces Seven Key European Consumer Trends for 2017." Mintel, November 3, 2016. http://www.mintel .com/press-centre/social-and-lifestyle/mintel-announces-seven-key -european-consumer-trends-for-2017.

Modern Salon. "2016 Professional Salon Industry Hair Care Study: Salon Services Grow 2.8%, Salon Hair Care Grows 3% in a Slow Economy." February 23, 2017. https://www.modernsalon.com/article/78630/2016 -professional-salon-industry-haircare-study-salon-services-grow-2-8 -salon-haircare-grows-3.

Monllos, Kristina. "Dove Moves away from Social Experiments in Its Latest Ad about Loving Your Hair." *Adweek*, April 21, 2016. http:// www.adweek.com/creativity/dove-moves-away-social-experiments-its -latest-ad-about-loving-your-hair-170976/.

Mukkanna, Krishna Sumanth, Natalie M. Stone, and John R. Ingram. "Para-phenylenediamine Allergy: Current Perspectives on Diagnosis and Management." *Journal of Asthma and Allergy* 2017, no. 10 (2017): 9–15. doi:10.2147/JAA.S90265.

Nelson, Elizabeth. "Hair Dyes and Relaxers Linked to Breast Cancer—What You Need to Know." The Breast Cancer Site. Accessed October 29, 2018. http://blog.thebreastcancersite.com/dyes-relaxers/.

Nielsen. "Green Generation: Millennials Say Sustainability Is a Shopping Priority." November 5, 2015. http://www.nielsen.com/us/en/insights/news/2015/green-generation-millennials-say-sustainability-is-a-shopping-priority.html?.

Nix, Joanna. "What the Heck Is 'Fragrance'? Thanks to California, Companies Now Have to Tell Us." *Mother Jones*, December 1, 2017. http://www.motherjones.com/politics/2017/12/what-the-heck-is-fragrance-thanks-to-california-companies-now-have-to-tell-us/#.

Norkin, Laura. "We Asked Women across the Country All about Their Hair." *InStyle*, August 6, 2018. https://www.instyle.com/beauty/splitting-hairs-survey-american-women-and-hair.

Oaklander, Mandy. "A New Fear about Food Dyes." *Prevention*, January 14, 2013. https://www.prevention.com/life/a20450128/blue-food-dyes-absorbed-into-bloodstream/.

Ocka, Ilissa. "9 Ways We Know Humans Triggered Climate Change." Environmental Defense Fund. Accessed October 31, 2018. https://www.edf.org/climate/9-ways-we-know-humans-triggered-climate-change.

Oram, Brian. "Hard Water Hardness Calcium Magnesium Water Corrosion Mineral Scale." Water Research Center. Accessed November 9, 2018. http://www.water-research.net/index.php/water-treatment/tools/hard-water-hardness.

Page, Shelley, and Susan Allen. "Ingredient in Men's Hair Dye Banned by Health Canada." November 8, 2015. https://www.pressreader.com.

Paton, Elizabeth. "Haircuts on a Global Scale." *New York Times*, July 20, 2016. https://www.nytimes.com/interactive/2016/07/20/fashion/global-haircut-prices.html.

Peet, Amanda. "Amanda Peet: Never Crossing the Botox Rubicon." *Lenny* (blog), April 26, 2016. https://www.lennyletter.com/story/amanda-peet-never-crossing-the-botox-rubicon.

Perception Institute. "The Good Hair Study." Accessed November 28, 2018. https://perception.org/goodhair/.

Personal Care Products Safety Act, S. 1014, 114th Congress (2015–2016). https://www.congress.gov/bill/114th-congress/senate-bill/1014?q=%7B%22search%22%3A%5B%22personal+care+products+safety+act%22%5D%7D&resultIndex=1.

Philipkoski, Kristen. "The Ultimate Beauty Luxury? Non-Toxic Color That Restores Your Pre-Gray Hair." *Forbes*, July 31, 2016. https://www.forbes

.com/sites/kristenphilipkoski/2016/07/31/the-ultimate-beauty-luxury
-non-toxic-color-that-restores-your-pre-gray-hair/#1d5a714f791a.

Pollan, Michael. *Food Rules: An Eater's Manual.* New York: Penguin Books,
2009.

Professional Consultants & Resources. "Beauty, Salon, Spa, and Cosmetics Industry Consulting, Research, and Reports." Accessed December 5, 2018. http://www.proconsultants.us.

Rachael. "Grey Hair Leads the Way for Beauty Trends in 2015." *HJi*, December 15, 2014. http://www.hji.co.uk/hair/grey-hair-beauty-trends-2015/.

Radico USA. *How to Apply.* Video, 1:29. Accessed November 9, 2018.
https://www.radicousa.com.

Reilly, Amanda. "Court Rebukes EPA, Orders Ban on Farm Chemical."
GreenWire, August 9, 2018. https://www.eenews.net/stories/1060093783.

Reuters. "Aging a Concern for Many Americans but Harder for Women:
Poll." *New York Daily News*, March 20, 2013. http://www.nydailynews
.com/life-style/health/aging-tougher-women-men-poll-article-1
.1294162.

Robin, Marci. "Everything You Need to Know about Organic and Natural Hair Color Formulas." *Good Housekeeping*, January 9, 2019. https://
www.goodhousekeeping.com/beauty-products/hair-dye-reviews/
advice/a17382/non-toxic-hair-dyes-55021302/.

Rosman, Katherine. "The Last Resort? Canyon Ranch Succumbs to Botox."
New York Times, December 2, 2017. https://www.nytimes.com/2017/12/02/
style/canyon-ranch-resort.html?em_pos=medium&emc=edit_li_20171202
&nl=nyt-living&nl_art=4&nlid=55963803&ref=headline&te=1&_r=0.

Rubin, Shira. "Your Obsession with 'Wellness' Is Killing the Dead Sea."
Daily Beast, November 27, 2017. https://www.thedailybeast.com/your
-obsession-with-wellness-is-killing-the-dead-sea.

Saitta, Peter, Christopher E. Cook, Jane L. Messina, Ronald Brancaccio,
Benedict C. Wu, Steven K. Grekin, and Jean Holland. "Is There a True
Concern regarding the Use of Hair Dye and Malignancy Development?"
Journal of Clinical and Aesthetic Dermatology 6, no. 1 (January 2013):
39–46. https://www.ncbi.nlm.nih.gov/pmc/articles/PMC3543291/.

Sarnoff, Rachel. "Henna How To: A Guide to Truly Natural Hair Col-
or." *Mommy Greenest* (blog), May 18, 2015. http://www.mommygreenest
.com/henna-how-to-a-guide-to-truly-natural-hair-color/.

Savers. "2017 Community Impact Report: A Reuseful Impact." https://www
.savers.com/sites/default/files/community_impact_report_2017-savers
.pdf.

Schneier, Matthew. "Fashion's Gaze Turned to Joan Didion in 2015." *New York Times*, December 18, 2015. https://www.nytimes.com/2015/12/20/fashion/joan-didion-celine-fashion-gaze.html.

Scranton, Alexandra. "Beauty and Its Beast." Women's Voices for the Earth, November 2014. https://womensvoices.org/wp-content/uploads/2014/11/Beauty-and-Its-Beast.pdf.

Scutti, Susan. "Group Sues FDA over Formaldehyde in Hair-Straightening Products." CNN, December 14, 2016. http://www.cnn.com/2016/12/14/health/hair-straightening-formaldehyde-fda/index.html.

Sensei, J. "Drama: Cutting Off One's Hair in Japan." Together with Japan, July 12, 2011. http://jp.learnoutlive.com/drama-cutting-off-ones-hair-in-japan/.

Sephora. "Hush." Accessed November 9, 2018. https://www.sephora.com/hush.

Shakespeare, Sebastian. "Sebastian Shakespeare: Nicky Clarke Claims Kate's Greying Hair Is a Disaster!" *Daily Mail*, July 28, 2015. http://www.dailymail.co.uk/femail/article-3178047/SEBASTIAN-SHAKESPEARE-Nicky-Clarke-claims-Kate-s-greying-hair-disaster.html.

Sokolove Law. "Just For Men Hair Color Dye Side Effects Lawsuits Explored." September 15, 2016. https://www.sokolovelawfirm.com/blog/just-for-men-dangers/.

Spurrier, Juliet, and BabyGearLab Team. "What Is Inside Those Disposable Diapers?" BabyGearLab, May 24, 2018. https://www.babygearlab.com/expert-advice/what-is-inside-those-disposable-diapers.

Sustainable Pulse. "French Safety Agency Discovers 60 Toxic Chemicals Including Glyphosate in Baby Diapers." January 23, 2019. https://sustainablepulse.com/2019/01/23/french-safety-agency-discovers-60-toxic-chemicals-including-glyphosate-in-baby-diapers/#.XEkUmy2ZP-a.

Sylvia Earle Alliance/Mission Blue. Accessed November 1, 2018. https://mission-blue.org.

Tawfik, Manal Said, and Hana BaAbdullah. "Migration Levels of Monostyrene in Most Vulnerable Foods Handled and Stored in Polystyrene Containers and Their Impact on the Daily Intake." Semantic Scholar, 2014. https://www.semanticscholar.org/paper/Migration-levels-of-monostyrene-in-most-vulnerable-Tawfik-BaAbdullah/fee7febac2da2215f319147c2b600251f2fa7cec.

Thomas, Robert McG. Jr. "Shirley Polykoff, 90, Ad Writer Whose Query Colored a Nation." *New York Times*, June 8, 1998. http://www.nytimes

.com/1998/06/08/nyregion/shirley-polykoff-90-ad-writer-whose
-query-colored-a-nation.html.

Tolentino, Jia. "How 'Empowerment' Became Something for Women to Buy." *New York Times Magazine*, April 12, 2016. https://www.nytimes .com/2016/04/17/magazine/how-empowerment-became-something -for-women-to-buy.html.

Tomlin, Annie. "The Beauty Side Effects You Don't Know About." Refinery29, August 27, 2014. http://www.refinery29.com/beauty-treatment -sickness-side-effects.

Trueman, Kerry. "'Drawdown' Gives Us 100 Uplifting Climate Solutions." Review of *Drawdown: The Most Comprehensive Plan Ever Proposed to Reverse Global Warming*, edited by Paul Hawken. Moms Clean Air Force, June 22, 2017. http://www.momscleanairforce.org/drawdown-book-review/.

Tunell, Alexandra. "#BeautySchool: The Real Difference between Natural, Organic, and Non-toxic Beauty Products." *Harper's Bazaar*, March 16, 2015. http://www.harpersbazaar.com/beauty/skin-care/advice/a10295/ natural-versus-organic-beauty-products/.

Unterberger, Lindsey. "Gray Hair, Don't Care: How Women Ditching Dye Could Uproot the Hair Care Industry." Beauty Independent, May 11, 2018. https://www.beautyindependent.com/gray-hair-color -dye-women-movement-acceptance-beauty-products-salons/.

US Department of Health and Human Services, National Toxicology Program. "Organization." Last modified October 31, 2018. https://ntp .niehs.nih.gov/about/org/index.html.

US Department of Labor, Occupational Safety and Health Administration. "Hazard Alert Update: Hair Smoothing Products That Could Release Formaldehyde." Accessed October 29, 2018. https://www.osha .gov/SLTC/formaldehyde/hazard_alert.html.

US Environmental Protection Agency. "Air Quality—National Summary." Last modified July 25, 2018. https://www.epa.gov/air-trends/air -quality-national-summary.

US Environmental Protection Agency. "Contaminants of Emerging Concern Including Pharmaceuticals and Personal Care Products." Last modified September 27, 2016. https://www.epa.gov/wqc/contaminants-emerging -concern-including-pharmaceuticals-and-personal-care-products.

US Environmental Protection Agency. "Fish and Shellfish Advisories and Safe Eating Guidelines." Last modified March 14, 2018. https://

www.epa.gov/choose-fish-and-shellfish-wisely/fish-and-shellfish
-advisories-and-safe-eating-guidelines.

US Food and Drug Administration. "Color Additives and Cosmetics." Last
modified June 19, 2018. https://www.fda.gov/forindustry/coloradditives/
coloradditivesinspecificproducts/incosmetics/ucm110032.htm.

US Food and Drug Administration. "Hair Dyes." Last modified October 30, 2018. https://www.fda.gov/cosmetics/productsingredients/
products/ucm143066.htm.

Vandenberg, Laura N., Theo Colborn, Tyrone B. Hayes, Jerrold J. Heindel,
David R. Jacobs Jr., Duk-Hee Lee, Toshi Shioda, et al. "Hormones
and Endocrine-Disrupting Chemicals: Low-Dose Effects and Non-
monotonic Dose Responses." *Endocrine Reviews* 33, no. 3 (June 1, 2012):
378–455. doi:10.1210/er.2011–1050.

Viégas, Fernanda, and Martin Wattenberg. "Wind Map." HintFM. Accessed November 1, 2018. http://hint.fm/wind/.

Visible Science. "How to Use Argan Oil for Gray Hair." Last modified
November 9, 2018. http://www.visiblesciencearganoil.com/how-to-use
-argan-oil-for-gray-hair.html.

Weil, Andrew. "Can Hair Dye Cause Cancer?" Weil: Andrew Weil,
M.D. Last modified August 17, 2017. https://www.drweil.com/health
-wellness/body-mind-spirit/cancer/can-hair-dye-cause-cancer/.

Weitz, Rose. *Rapunzel's Daughters: What Women's Hair Tells Us about Wom-
en's Lives.* New York: Farrar, Straus and Giroux, 2004.

Westervelt, Amy. "How Legislation Potentially Affects Harmful Chemicals in Your Beauty Products." *Teen Vogue*, November 9, 2017. https://
www.teenvogue.com/story/harmful-chemicals-in-beauty-products
-legislation-guide.

White, Kyle. "How to Make Your Hair Color Last *Way* Longer." Re-
finery29, October 17, 2014. https://www.refinery29.com/hair-color
-maintenance.

Wiseman, Rebecca. "How to Remove Yellow from Gray Hair." Love to
Know. Accessed November 1, 2018. https://seniors.lovetoknow.com/
Remove_Yellow_from_Gray_Hair.

Wolfe, David. "How Climate Change Affects the Monarch Butterfly, and
What We Can Do about It." Environmental Defense Fund, May 26,
2016. https://www.edf.org/blog/2016/05/26/how-climate-change-affects
-monarch-butterfly-and-what-we-can-do-about-it.

Women's College Hospital, Women's Health Matters. "Environmental Health." Accessed November 2, 2018. http://www.womenshealthmatters .ca/health-centres/environmental-health/.

Women's Voices for the Earth. "Beauty and Its Beast: Unmasking the Impacts of Toxic Chemicals on Salon Workers: Fact Sheet." Accessed October 29, 2018. https://www.womensvoices.org/safe-salons/beauty -and-its-beast/.

Women's Voices for the Earth. "Report Exposes Industry-Funded Cosmetics Ingredient Review (CIR) Panel's Failure to Protect the Public and Manufacturers." April 24, 2018. https://www.womensvoices.org/ 2018/04/24/report-exposes-industry-funded-cosmetics-ingredient -review-cir-panels-failure-protect-public-manufacturers/.

Women's Voices for the Earth. "Women, Health, and the Environment." Accessed October 29, 2018. https://www.womensvoices.org/about/why -a-womens-organization/.

WS. "Why Does White (or Gray, Light Blonde, Highlighted) Hair Turn Yellow: And What to Do about It." *Science-y Hair Blog*, February 7, 2015. https://science-yhairblog.blogspot.com/2015/02/why-does-white -or-gray-light-blonde.html.

Yoder, Kate. "The EPA Is Making 'Transparency' Look a Helluva Lot Like Censorship." Grist, March 27, 2018. https://grist.org/briefly/the-epa -is-making-transparency-look-a-helluva-lot-like-censorship/?utm _medium=email&utm_source=newsletter&utm_campaign=weekly.

Island Press | Board of Directors